AF594940

Forests as Nature-Based Solutions: Ecosystem Services, Multiple Benefits and Trade-Offs

Forests as Nature-Based Solutions: Ecosystem Services, Multiple Benefits and Trade-Offs

Editors

Elisabetta Salvatori
Giacomo Pallante

MDPI • Basel • Beijing • Wuhan • Barcelona • Belgrade • Manchester • Tokyo • Cluj • Tianjin

Editors
Elisabetta Salvatori
Department for Sustainability, Italian National Agency for New Technologies, Energy and Sustainable Economic Development (ENEA)
Italy

Giacomo Pallante
Department for Sustainability, Italian National Agency for New Technologies, Energy and Sustainable Economic Development (ENEA)
Italy

Editorial Office
MDPI
St. Alban-Anlage 66
4052 Basel, Switzerland

This is a reprint of articles from the Special Issue published online in the open access journal *Forests* (ISSN 1999-4907) (available at: https://www.mdpi.com/journal/forests/special_issues/naturebased_solutions).

For citation purposes, cite each article independently as indicated on the article page online and as indicated below:

LastName, A.A.; LastName, B.B.; LastName, C.C. Article Title. *Journal Name* **Year**, *Volume Number*, Page Range.

ISBN 978-3-0365-4975-0 (Hbk)
ISBN 978-3-0365-4976-7 (PDF)

Cover image courtesy of Elisabetta Salvatori

Contents

About the Editors

Elisabetta Salvatori

Elisabetta Salvatori graduated in biology with a Ph.D. in Ecology. She is a researcher at the Italian National Agency for New Technologies, Energy and Sustainable Economic Development (ENEA), Department for Sustainability. Her main research interests concern the biophysical quantification of Ecosystem Services, particularly of regulating ES provided by urban and peri-urban Green Infrastructures, and the use of nature-based solutions for urban and territorial regeneration. She has also worked on the characterization of plant functional traits in response to environmental stress factors (air pollution, drought, salt stress, nitrogen deposition), on the study of phytoremediation of pollutants (heavy metals, anionic surfactants) in contaminated soils and waters, and on the investigation of environmental quality through bioindication and biomonitoring. She is a member of the Editorial Board of Environmental and Experimental Botany and Forests. Since 2006, she has been a member of the Italian Society of Ecology (SItE). She has authored more than 40 scientific publications in international journals with impact factors.

Giacomo Pallante

Giacomo Pallante is a researcher at the department for Sustainability of ENEA and a lecturer of Applied Economics at the University of Rome Tor Vergata. Giacomo's research interests are in environmental economics and development economics. He has co-authored articles published in international journals such as Ecological Economics, World Development and Food Policy. He completed a Master's in economics and a Ph.D. in environmental economics at the University of Rome Tor Vergata in 2011 and 2014, respectively. Prior to joining ENEA, he was an environmental economist at the Italian Ministry of Environment, Italian delegate at the OECD Working Party on Environmental Performance, and a research fellow at the University of Rome Tor Vergata. He held a position as a lecturer of Environmental Economics at John Cabot University and Economic Policy at the University of Rome Tor Vergata.

Preface to "Forests as Nature-Based Solutions: Ecosystem Services, Multiple Benefits and Trade-Offs"

This collection of papers presents relevant results from scientific researchers about the ecosystem services provided by forests as nature-based solutions in natural and urban contexts, encompassing not only provisioning services, but also regulation and maintenance services, such as carbon and air pollution sink, as well as recreational services. The impacts of environmental changes on forest multifunctionality and provision of services are also investigated. Case studies for monetary valuation, willingness to pay for ecosystem services, and cost/benefit analyses are discussed, as well as the potential trade-offs and synergies between services, which might result from different stakeholders' perspective and management strategies.

Elisabetta Salvatori and Giacomo Pallante
Editors

Editorial

Forests as Nature-Based Solutions: Ecosystem Services, Multiple Benefits and Trade-Offs

Elisabetta Salvatori * and Giacomo Pallante

Department for Sustainability (SSPT-STS), Italian National Agency for New Technologies, Energy and Sustainable Economic Development (ENEA), R.C. Casaccia, Via Anguillarese, 301 Santa Maria di Galeria, 5-00123 Rome, Italy; giacomo.pallante@enea.it
* Correspondence: elisabetta.salvatori@enea.it; Tel.: +39-06-30483164

Keywords: climate change mitigation and adaptation; air quality; water quality; recreation; plant functional traits; land-use planning; forest management and restoration; protected areas; monetary valuation and accounting; socio-economic benefits

Citation: Salvatori, E.; Pallante, G. Forests as Nature-Based Solutions: Ecosystem Services, Multiple Benefits and Trade-Offs. *Forests* **2021**, *12*, 800. https://doi.org/10.3390/f12060800

Received: 7 June 2021
Accepted: 16 June 2021
Published: 18 June 2021

Forest ecosystems, including natural forests, managed forests, agroforestry systems, and urban and peri-urban forests, can be considered as multifunctional Nature-based Solutions (NbS) since they deliver key ecosystem services to people. The concept of NbS is an "umbrella" framework for several ecosystem-based approaches, categorized by the IUCN as protective (e.g., area-based conservation), restorative (e.g., ecological restoration), infrastructure-based (e.g., green infrastructure), management-based, or issue-specific (e.g., ecosystem-based disaster risk reduction) [1]. All of these approaches rely on biodiversity and ecosystem services to address global societal challenges, simultaneously providing environmental, social and economic benefits, and helping communities build resilience [2].

NbS are becoming more and more relevant in international and European policy frameworks, such as in the EU Biodiversity Strategy 2030 [3] and in the upcoming EU Forest Strategy. However, for the effective implementation and mainstreaming of NbS, several research gaps still need to be addressed.

These include the need for collecting further evidence about the ecosystem services provided by forests in natural, semi-natural and urban contexts, encompassing not only provisioning services (e.g., timber, raw materials) but also regulation and maintenance, as well as cultural services. Indeed, forests are fundamental for climate regulation, carbon sequestration, air, soil, and water quality improvement, and for mitigating natural hazards, providing also recreation, spiritual enrichment and aesthetic experience, that contribute to human wellbeing. Key ecological characteristics of forests (e.g., plant functional traits [4]) supporting the delivery of multiple benefits, as well as the possible impacts of climate change on forest functionality and services provision, should also be further investigated.

Another research priority concerns the assessment of the trade-offs between services which might result from different stakeholders' objective function and management strategies [5]. Monetary valuation, ecosystem services accounting, and cost-benefit analysis are intended to inform citizens, firms, and policy makers about the contribution of forests to private and public benefits and the welfare consequences of alternative forest planning [6]. In this context, the potential disservices, as well as the limitations in using forests as NbS for addressing specific challenges, must be identified and critically discussed by adopting a science-policy interface approach [7].

This special issue entitled "Forests as Nature-Based Solutions: ecosystem services, multiple benefits and trade-offs" encourages studies that deal with the above-mentioned ecological, economic, and social aspects. The aim is to stimulate discussion between scientists and to propose solutions for the operationalization of forests as NbS, thus supporting stakeholders in decision-making processes.

Funding: This research received no external funding.

Data Availability Statement: Not applicable.

Conflicts of Interest: The authors declare no conflict of interest.

References

1. Cohen-Shacham, E.; Walters, G.; Janzen, C.; Maginnis, S. *Nature-Based Solutions to Address Societal Challenges*; International Union for Conservation of Nature: Gland, Switzerland, 2016. [CrossRef]
2. European Commission. Nature-Based Solutions. Available online: https://ec.europa.eu/info/research-and-innovation/research-area/environment/nature-based-solutions_en (accessed on 20 May 2021).
3. European Commission. Communication from the Commission to the European Parliament, The Council, The European Economic and Social Committee and The Committee of the Regions. EU Biodiversity Strategy for 2030. Available online: https://eurlex.europa.eu/resource.html?uri=cellar:a3c806a6-9ab3-11ea-9d2d-01aa75ed71a1.0001.02/DOC_1&format=PDF (accessed on 20 May 2021).
4. Hanisch, M.; Schweiger, O.; Cord, A.F.; Volk, M.; Knapp, S. Plant functional traits shape multiple ecosystem services, their trade-offs and synergies in grasslands. *J. Appl. Ecol.* **2020**, *57*, 1535–1550. [CrossRef]
5. Grace, M.; Balzan, M.; Collier, M.; Geneletti, D.; Tomaskinova, J.; Abela, R.; Borg, D.; Buhagiar, G.; Camilleri, L.; Cardona, M.; et al. Priority knowledge needs for implementing nature-based solutions in the Mediterranean islands. *Environ. Sci. Policy* **2021**, *116*, 56–68. [CrossRef]
6. Hein, L.; Bagstad, K.J.; Obst, C.; Edens, B.; Schenau, S.; Castillo, G.; Soulard, F.; Brown, C.; Driver, A.; Bordt, M.; et al. Progress in natural capital accounting for ecosystems. *Science* **2020**, *367*, 514–515. [CrossRef] [PubMed]
7. Kieslich, M.; Salles, J.M. Implementation context and science-policy interfaces: Implications for the economic valuation of ecosystem services. *Ecol. Econ.* **2021**, *179*, 106857. [CrossRef]

MDPI

Article

Facing Multiple Environmental Challenges through Maximizing the Co-Benefits of Nature-Based Solutions at a National Scale in Italy

Elena Di Pirro [1], Lorenzo Sallustio [1], Joana A. C. Castellar [2,3], Gregorio Sgrigna [4], Marco Marchetti [1,*] and Bruno Lasserre [1]

[1] Dipartimento di Bioscienze e Territorio, Università degli Studi del Molise, 86090 Pesche (IS), Italy; elena.dipirro@unimol.it (E.D.P.); lorenzo.sallustio@unimol.it (L.S.); lasserre@unimol.it (B.L.)
[2] University of Girona, Plaça de Sant Domènec 3, 17004 Girona, Spain; jcastellar@icra.cat
[3] Catalan Institute for Water Research (ICRA), Carrer Emili Grahit 101, 17003 Girona, Spain
[4] Consiglio Nazionale delle Ricerche (CNR), Istituto di Ricerca sugli Ecosistemi Terrestri (IRET), URT UniSalento, Centro Ecotekne, Monteroni, Pal. B-S.P. 6 Lecce, Italy; gregorio.sgrigna@iret.cnr.it
* Correspondence: marchettimarco@unimol.it

Abstract: The European Union is significantly investing in the Green Deal that introduces measures to guide Member States to face sustainability and health challenges, especially employing Nature-Based Solutions (NBS) in urban contexts. National governments need to develop appropriate strategies to coordinate local projects, face multiple challenges, and maximize NBS effectiveness. This paper aims to introduce a replicable methodology to integrate NBS into a multi-scale planning process to maximize their cost–benefits. Using Italy as a case study, we mapped three environmental challenges nationwide related to climate change and air pollution, identifying spatial groups of their co-occurrences. These groups serve as functional areas where 24 NBS were ranked for their ecosystem services supply and land cover. The results show eight different spatial groups, with 6% of the national territory showing no challenge, with 42% showing multiple challenges combined simultaneously. Seven NBS were high-performing in all groups: five implementable in permeable land covers (urban forests, infiltration basins, green corridors, large parks, heritage gardens), and two in impervious ones (intensive, semi-intensive green roofs). This work provides a strategic vision at the national scale to quantify and orient budget allocation, while on a municipal scale, the NBS ranking acts as a guideline for specific planning activities based on local issues.

Keywords: human health; human well-being; urban sustainability; green deal; urban forests; green roofs; multifunctionality

Citation: Di Pirro, E.; Sallustio, L.; Castellar, J.A.C.; Sgrigna, G.; Marchetti, M.; Lasserre, B. Facing Multiple Environmental Challenges through Maximizing the Co-Benefits of Nature-Based Solutions at a National Scale in Italy. *Forests* **2022**, *13*, 548. https://doi.org/10.3390/f13040548

Academic Editors: Elisabetta Salvatori and Giacomo Pallante

Received: 24 February 2022
Accepted: 29 March 2022
Published: 31 March 2022

1. Introduction

In the European Union (EU), air pollution and the extreme events related to climate change (e.g., heatwaves and floods) are exerting pressure both on human health and natural capital integrity [1], leading to millions of premature deaths and economic losses each year [2]. This is especially relevant in urban areas, where 73% of the European population lives, compared to 50% globally [3,4]. For this reason, the EU is significantly investing in the European Green Deal, which introduces legislative and non-legislative measures to legally bind and guide the Member States to face sustainability and health challenges. The EU fixed targets across different strategies (e.g., Forestry and Biodiversity Strategy to 2030), laws (e.g., European climate law), and action plans (e.g., zero pollution action plan) [5] that the Member States need to meet at the national level for improving the quality of ecosystems and human life [6]. For example, a recent study by Khomenko et al. [7] estimated that about 52,000 lives would be saved annually if 1000 European cities met World Health Organization (WHO) air-quality standards. Particular attention is paid to policies and planning at the local scale to reconfigure urban areas so that they consume fewer resources, generate less pollution (including greenhouse gases), and become more resilient and sustainable [4] while facing budgetary pressure [8].

As a consequence, there is a growing interest in valuing Ecosystem Services (ES) and including them in decision-making processes [1,9] as a lens to achieve environmental and societal goals [10]. Hence, the concept of Nature-Based Solutions (NBS) rises as an environmentally friendly alternative to favor the provision/maintenance of ES. NBS are defined as "solutions that are inspired and supported by nature, which are cost-effective, simultaneously provide environmental, social and economic benefits and help build resilience" [2]. NBS is an umbrella concept related to and integrated into other concepts, such as green and blue infrastructure, urban forestry, ecological engineering, disaster risk reduction, and ecosystem-based adaptation [11–14]. These concepts were introduced to address the challenges from distinct perspectives while the NBS strength is the integrated perspective for providing co-benefits and generating win–win solutions (i.e., multifunctionality) [13,15,16]. Moreover, implementing NBS can foster both human well-being and biodiversity cost-effectively while offering new job and innovation opportunities [17].

Therefore, both governmental and non-governmental organizations are offering huge funding globally [18,19] to enable the implementation of NBS [20,21]. The focus is predominantly on afforestation and reforestation programs [22], e.g., "3 billion trees" in the EU [23], the "trillion tree campaign" [24], and the "great green wall" [25]. Notwithstanding, McDonald et al. [26] highlight that funds for tree-planting and maintenance initiatives are often constrained or limited by planning silos. Indeed, governments generally receive several indications on NBS from the EU that are not easily translated into effective and practical urban greening programs. For this purpose, the Horizon 2020 program classified NBS as a priority area of investment to enhance the resilience in urban areas in the face of global changes [1] and establish Europe as a world leader of NBS [10,27]. Demonstration projects of NBS—and related concepts—are in place in several cities of the Member States to tackle different urban issues such as the mitigation of air pollution, temperature extremes, noise, drought, and flooding [27–30]. EU-funded projects on NBS work in task forces to improve knowledge, reduce duplication, and facilitate progress towards shared goals [31]. These projects are proving to be a catalyst for research-practice partnerships [18], increasing knowledge and awareness regarding NBS indicators, impacts, performance, and cost-effectiveness assessment, building repositories with different case studies (e.g., OPPLA, the online EU repository of NBS). All projects aim at strengthening NBS regional development and translating results from experts to stakeholders [6]. More details regarding the status of H2020 projects are available from Wild et al. [31].

However, projects are still often implemented as standalone experiments in urban areas, scattered and uncoordinated throughout various policy levels and sectors [32,33]. As hubs of population and socio-economic activity, urban areas represent concentrated opportunities for addressing issues of sustainability at the local scale [4,18]. Nevertheless, ameliorating the environmental conditions in a few cities can only partially contribute to delivering the national-level commitments that countries have with the EU and with United Nations [33]. Therefore, lessons learned from single case studies need to be coordinated across multiple political and geographical levels to enable the long-term respect of national targets and international commitments [34,35]. Despite the fact that NBS are implicitly or explicitly cited in different European legislative frameworks [6,36,37], the H2020 NATURVATION project underlined as a legal initiative or policy coordination at the EU level requiring Member States to systematically program and invest in NBS is still absent [32]. In addition, the review conducted by Mendonça et al. [20] reveals that the policy instruments to mainstream the NBS concept into policy are usually investigated just at the city level, thus neglecting the country or higher levels of implementation.

However, considering the huge funding opportunity for the member states envisaged by the EU Green Deal—and other key policy initiatives [37]—national governments need to develop appropriate strategies to coordinate local projects and face multiple and complex challenges throughout the territory [33] to maximize the NBS effectiveness [22,32] and involve all relevant stakeholders [9,38]. Although this strategic level is still missing in most of the member states [32], it is crucial, especially for countries located in vulnerable areas currently facing climate and pollution issues (e.g., Mediterranean region [39]). In these countries, a wide and national perspective could help to coordinate the implementation of NBS at lower levels for reaching multiple national targets related to different environmental policies, with the final scope to improve the state of ecosystems and human health as a whole. Accordingly, we selected Italy as a case study, since it is a representative member state both for the challenges related to pollution and climate change and for the national policies in place to improve urban sustainability.

This paper thus aims to introduce a replicable methodology to integrate and strengthen NBS into a multi-level planning process to maximize their cost–benefit from the large-scale policy and planning initiatives (e.g., national) to the local scale (e.g., municipal). Generally, the former is focused on ameliorating environmental sustainability through reaching fixed targets, while the second is oriented to directly reconfigure urban areas, improving the wellbeing of inhabitants. Although many authors have already dealt with methods and approaches for planning and designing multifunctional NBS [40,41], they are usually limited to the municipal scale. Therefore, to the best of our knowledge, this is the first framework conducted on a national scale. We started from the Environmental Quality Standards set by Di Pirro et al. [42] to map the spatial co-occurrence of the same environmental challenges (individual or multiple) nationwide (i.e., spatial groups). These spatial groups of challenges serve as functional areas where NBS providing multiple co-benefits can be identified to effectively mitigate peculiar multiple environmental stressors altering human health and well-being (i.e., air pollution, heatwaves, flood hazards). Consequently, different rankings of 24 NBS were built for each spatial group based on i) their capacity to supply ES, and hence their performance to address the challenges, and ii) the current land cover in which they can be implemented. Using the Environmental Quality Standards helped to establish a replicable, clear, and spatially explicit understanding of the challenges that the country needs to tackle. The here-proposed framework is able to support the strategical coordination of national funds allocation and to enhance the effectiveness of interventions at the local scale consistent with the national objectives.

2. Case Study

In the Mediterranean basin, climate change has exacerbated existing environmental challenges caused by the combination of increasing pollution, land use changes, and declining biodiversity [39]. Indeed, Italy is consistently experiencing the adverse effects of climate change, such as heatwaves, floods, and drought events, combined with the strong exposition of the three most harmful air pollutants in the EU [39,43–46]. In addition to these challenges, within the Italian territory, the sealed surface reaches one of the highest relative national coverages (7.1% [47]) among EU countries [48]. The scattered and fragmented urban mosaic [49] has smoothed the boundaries between urban and rural areas [50], exacerbating issues related to ecological connectivity, biodiversity, and ecosystem services loss [51–54]. Therefore, the Italian national government envisaged different urban sustainability strategies and policies and set ambitious tree-planting objectives, based on the premise that planning urban forests is a feasible response to current challenges and that they can enhance the resilience of cities and safeguard the population's health [55]. For example, the "Decree on Climate" [56] is a national policy adopted to tackle the climate emergency and achieve the objectives related to the EU Air Quality Directive [57]. Within the decree, the "Urban Forestry Program" (Azioni per la riforestazione-Art.4 [56]) allocates funds to implement urban and peri-urban forests and to reduce impervious surfaces (i.e., de-sealing actions), as key interventions to address urban challenges [58]. The National Strategy on Urban Greenspaces [59] guides the municipalities to the effective implementation of local-scale initiatives for strengthening ecological networks. The Urban Forestry Program allocates funds to just 14 metropolitan cities, while the National Strategy on Urban Greenspaces includes all the Italian municipalities in its analysis. However, Di Pirro et al. [42] reveal an incomplete spatial agreement between the current fund's allocation envisaged by the Urban Forestry Program and the real spatial distribution of the challenges to address. Sallustio et al. [60] highlight that the inclusion of all municipalities can ensure an equitable distribution of economic resources and provide guidelines that can be easily replicated and implemented at the municipal scale.

While the current policy mix provides a starting point to promote/maintain NBS, there is significant potential on the national level to uptake NBS into policy and optimize the rationale of budget allocation to design an optimized NBS network. Accordingly, we provide a wide strategic perspective that can support the allocation of funds currently envisaged by the EU Green Deal. We include the whole national territory in the identification, first of the challenges' distribution and then in the most effective and multifunctional NBS available for their mitigation.

3. Materials and Methods

This study was developed according to the three stages shown in Figure 1. Stage I: the identification and mapping of three environmental challenges in Italy (i.e., air quality, climate adaptation and mitigation, and water management), adopting the Environmental Quality Standards proposed by Di Pirro et al. [42]; Stage II: the overlay of the three challenges allows the identification of portions of territory threatened simultaneously by the same challenges (i.e., spatial groups); Stage III: a ranking of 24 NBS suitable to address the challenge(s) for each spatial group is proposed, based on the NBS performance assessment provided by Castellar et al. [30] and the land cover (Figure 1). All the analyses are conducted by a pixel-based approach.

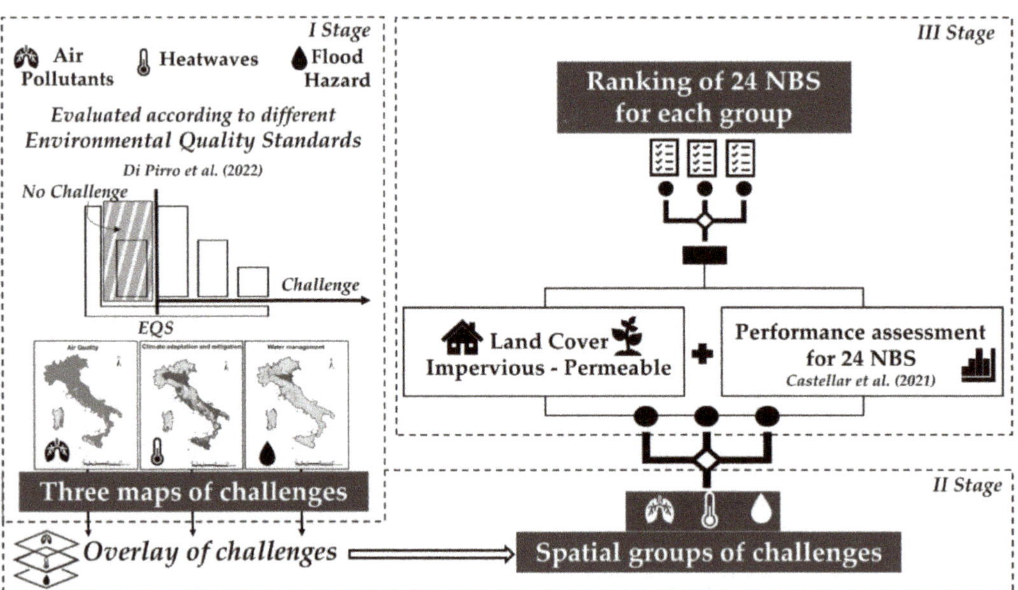

Figure 1. Workflow developed according to three main stages.

3.1. *Environmental Challenges in Italy and Their Combination in Spatial Groups*

According to Stage I (Figure 1), we considered three challenges, air quality, climate adaptation and mitigation, and water management (following Raymond et al. [15]), defined by the presence of three environmental stressors (air pollutants, frequency of heatwaves, flood hazard, respectively) altering human health when they exceed specific thresholds [61] (e.g., Environmental Quality Standards—EQS [62]). Consequently, to identify the portion of national territory threatened by these challenges, we adopted the methodological framework proposed by Di Pirro et al. [42], where different EQS were selected and used as common thresholds to assess environmental and societal demands. The EQS proposed by Di Pirro et al. [42] are defined according to (i) the European standards set in Air Quality Directive (2008/50/EC), (ii) the definition of heatwaves and projections of climate change given by [43], and (iii) the flood hazard estimated by ISPRA [63] (further details are reported in [42]).

To define portions of the Italian territory showing air quality challenge, we considered EQS for the three most harmful pollutants in the EU, namely PM_{10}, NO_2, and O_3 [57]. All the pixels showing at least one of the three pollutants in exceedance for the respective EQS are considered as portions of territory where air quality regulation is needed to address the challenge. As regards the challenge of climate adaptation and mitigation, we considered the EQS of 4 days/year of heatwaves [43]. Hence, all the pixels exceeding this EQS are considered as portions of territory where climate regulation is needed to address the challenge. Finally, for the water management challenge, we considered EQS based on the probability that a flooding event will occur, according to estimates of their return period [63]. Hence, all the pixels in flood hazard are considered as portions of territory where water regulation is needed to address the challenge. Thus, starting from the methodological framework proposed by Di Pirro et al. [42], we derived three maps with the spatial distribution of the three challenges, with a spatial resolution of 1 km, as shown in Figure 2.

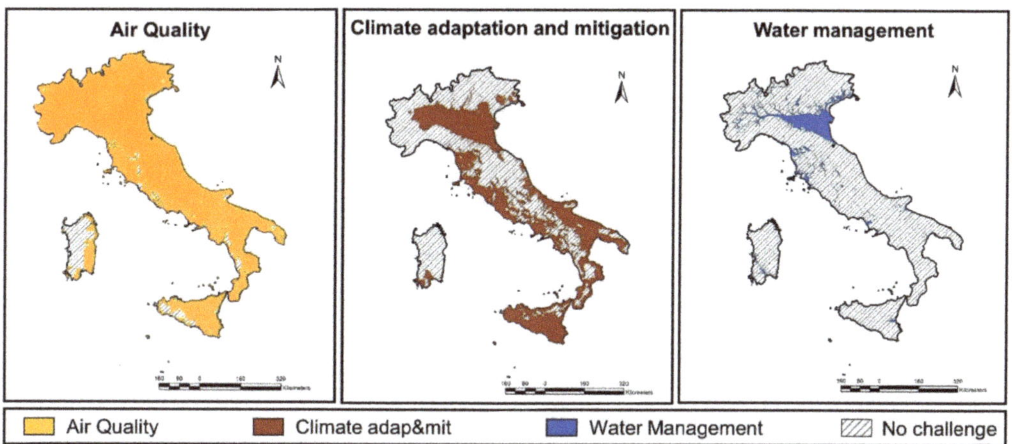

Figure 2. The three challenges considered: air quality, climate adaptation and mitigation, and water management. The areas with no challenge are represented by the striped pattern.

According to Stage II (Figure 1), the three maps of challenges were then combined in a GIS environment to investigate where, which, and how many challenges overlay in the same portion of the territory, thus needing to be addressed simultaneously. From this analysis, we obtained different homogenous groups, where interventions need to be differentiated to address the specific challenge(s). According to Stage III (Figure 1), for each group, we explored (i) the population density [64], to estimate the inhabitants exposed to single or multiple challenges as well as the potential beneficiaries of NBS; and (ii) the land cover, to define quantity and typology of space available for NBS implementation. We focused our analysis on two land covers (i.e., impervious and permeable) using the 2018 High-Resolution Layers, with a spatial resolution of 20 m [65]. The impervious surfaces were reclassified according to Congedo et al. [66], thus considering values of the Degree of

Imperviousness greater than 29%. The permanent water bodies (covering about 1.4% of the national territory) were neglected, since specific policies (e.g., water quality and security) and NBS might be implemented, out of the scope of this work.

3.2. Calculating the Nature Based Solutions Performance in Dealing with Challenges

Relying on the NBS capacity to provide multiple ES and mitigate environmental stressors, we assumed that NBS are the unique interventions to address the challenges in each group. In the literature, NBS encompasses a wide range of interventions and actions. Following the classification proposed by Eggermont et al. [67], NBS can be considered as conservation and restoration of ecosystems (i.e., Type 1), sustainable management for improving ES supply (i.e., Type 2), and the creation of new ecosystems (i.e., Type 3). For this work, we considered only the creation of new ecosystems, i.e., NBS Type 3 according to Eggermont et al. [67], focusing in particular on NBS spatial and technological units proposed by Castellar et al. [30]. New NBS thus need to be identified and differentiated according to their capacity to provide ES able to address the challenge(s) (i.e., performance). We adopted the performance assessment proposed by Castellar et al. [30], where for 32 NBS they calculated a performance score (PS), ranging from 0 to 1, representing the NBS capacity to provide ES able to address ten different challenges. We limited our analysis to 24 NBS and respective PS related to the three challenges under investigation in this study (i.e., air quality, climate adaptation and mitigation, and water management). Eight NBS were thus excluded since they are not terrestrial or not considered as Type 3 (i.e., new ecosystem). Therefore, when a single challenge characterizes the group, we reported the same PS of Castellar et al. [30]; when multiple challenges characterize the group, we averaged the PS for the respective challenges of the group. Accordingly, starting from the 24 NBS, we produced different rankings of PS as many as the groups of challenges, thus allowing us to select the best performing set of NBS to effectively address the environmental challenges.

3.3. Classification of Nature Based Solutions for Land Covers

The 24 NBS considered for this study are the following: community gardens, constructed wetlands, extensive green roofs, green corridors, green façades, green wall systems, heritage gardens, infiltration basins, intensive green roofs, large urban parks, planter green walls, pocket gardens/parks, private gardens, raingardens, semi-intensive green roofs, street trees, swales, urban forests, urban orchards, vegetated grid paves, vegetated pergolas, vertical mobile gardens, (wet) retention ponds, and shelters for biodiversity.

We classified NBS into two categories: implementable in impervious surfaces (I-NBS) and permeable surfaces (P-NBS). The classification is based on descriptions provided by Castellar et al. [30], considering both the NBS' structures and sizes. Accordingly, I-NBS are those implementable on buildings (i.e., green façades, green wall systems, vertical mobile gardens, planter green walls, vegetated pergolas, extensive green roofs, intensive green roofs, semi-intensive green roofs), and along streets and parking lots, close to buildings and houses (i.e., raingardens, swales, street trees, and vegetated grid paves, pocket gardens/parks, private gardens). Except for vegetated grid paves, where we consider the conversion from traditional car parks to green parking lots, we excluded the possibility of de-sealing actions (land cover changes from impervious to permeable), e.g., building's removal and conversion to a large urban park. On the other side, P-NBS are those that can be implemented on permeable land covers (green corridors, large urban parks, urban forests, heritage gardens, community gardens, urban orchards, infiltration basins, (wet) retention ponds, constructed wetlands). NBS similar to each other for the structure but not for the size (e.g., pocket gardens vs large parks) were distinguished by a 0.5 ha threshold [30]. Consistently with the HRL spatial resolution used to estimate land covers surfaces [65], we reduced this dimensional threshold to 400 m^2 (i.e., one pixel). In this way, pocket gardens were assigned to I-NBS while large parks were assigned to P-NBS.

Therefore, for each group, we can provide the quantity and typology of challenges to address, the incidence of permeable and impervious land covers, a ranking of P-NBS and I-NBS ranging from 0 to 1 based on their ability to address the specific challenges of the group.

4. Results

Eight different spatial groups resulted from the spatial combination of the three challenges; Figure 3 shows the map with their spatial distribution while the pie-chart shows the relative coverage of each spatial group. Only 6% of the national territory shows no challenge ("NoChal" group). Conversely, 7.9% of the territory shows all the three challenges combined simultaneously ("ALL" group). Three groups show the individual challenge covering 51.5% of the national territory: 47% air quality ("AIR" group), 4.3% climate adaptation and mitigation ("CLIM" group), and 0.2% water management flood hazard ("WAT" group). The remaining 35% of the territory is occupied by the last three groups, characterized by two challenges simultaneously. The challenge of air quality co-occurring with climate adaptation and mitigation covers 33% of the national territory ("AIR-CLIM" group), while its combination to water management spans over 1.2% of the national territory ("AIR-WAT" group). Finally, when the spatial combination is between climate adaptation and mitigation and water management, the group covers 0.3% of the country ("CLIM-WAT" group).

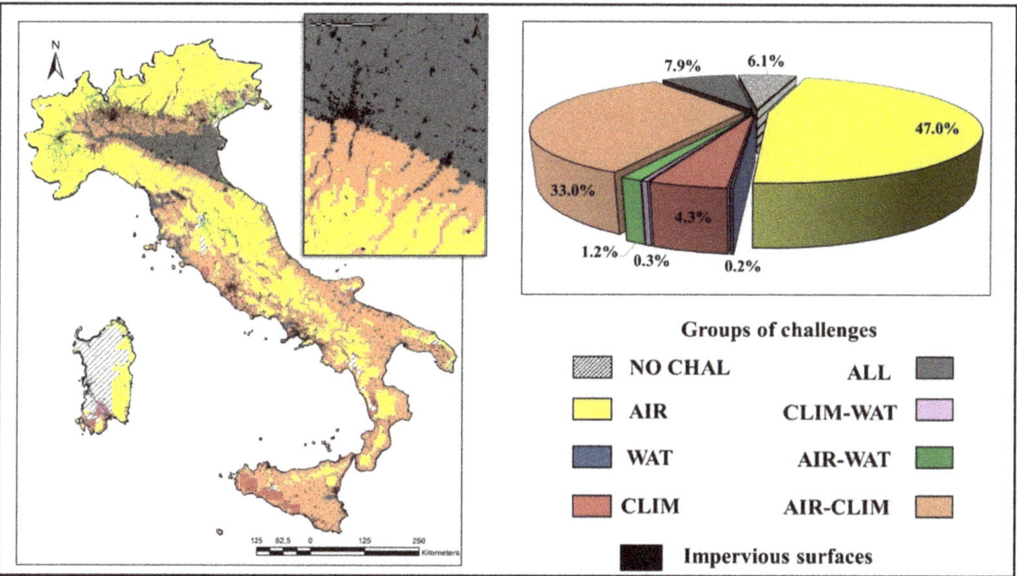

Figure 3. The eight spatial groups of challenges. The pie chart shows the relative coverage of each spatial group (% of the national territory). In black on the map is shown the spatial distribution of impervious surfaces throughout the national territory.

Built-up areas in Italy cover about 7.1% of the national territory [47], and their incidence is quite variable across the groups (Table 1 and Figure 3). The AIR and the NoChal groups are the only ones showing a relatively impervious surface lower than the national one (3% and 4%) as well as the lowest population density (91 and 150 inhab/km^2). On the contrary, CLIM and CLIM-WAT groups show the highest relative impervious surfaces (18% and 24% respectively) as well as the highest population density (1036 and 1086 inhab/km^2). AIR-WAT, ALL, WAT, and AIR-CLIM groups, respectively, show the following relative impervious surfaces, slightly higher than the national one: 13%, 11%, 9%, and 8%, with intermediate values of population density, 326, 257, 254, and 213 inhab/km^2.

Table 1. For each group, the area in km^2, coverage of permeable and impervious surface (km2), number of inhabitants, and population density (inhab/km2) within the groups are reported.

Groups	Area (km^2)	Permeable (km^2)	Impervious (km^2)	Population (n° inhab)	Pop Dens (inhab/km^2)
AIR	141,044	136,310	4734	12,974,163	91
CLIM	12,582	10,345	2237	13,468,447	1036
WAT	477	435	42	130,275	254
AIR-CLIM	97,769	89,603	8166	21,377,514	213
AIR-WAT	3352	2904	448	1,238,194	326
CLIM-WAT	1043	790	253	1,152,988	1086
ALL	22,875	20,446	2429	6,163,604	258
NoChal	18,393	17,639	754	2,802,590	150

All the 24 NBS show Performance Scores (PS) ranging from the minimum value of 0, only in the groups characterized by single challenges, to the maximum of 1, in each group. We divided PS into three classes, low PS (0–0.33), medium PS (0.33–0.66), high PS (0.66–1), to more facilitate the reading of the different performances.

The number of NBS with high PS (Table 2) is variable across the groups ranging from 16 in the WAT group to 11 in AIR-WAT. NBS with high PS can be implemented in both permeable (with a maximum of 9 P-NBS in the WAT group) and impervious land covers (with a maximum of 9 I-NBS in the CLIM group). P-NBS have similar PS and ranking among the different groups, while we registered more dissimilarity among the I-NBS both for PS values and ranking. This difference is particularly marked for I-NBS in AIR and WAT groups. Accordingly, their combination in the AIR-WAT group shows the least number of high PS (4 I-NBS), highlighting a lack of synergy between ES for simultaneously addressing the challenges of air quality and water management.

Vertical green (i.e., green façades, green wall systems, vegetated pergola, vertical mobile gardens) shows high PS for the mitigation of both air pollutants and heatwaves while low PS for the mitigation of flood hazard. We found the opposite PS trend for rain gardens, swales, and vegetated grid paves, which are particularly useful to mitigate flood hazards and so address the challenge of water management.

Seven NBS have high PS simultaneously in all groups: five P-NBS (i.e., infiltration basin, green corridors, urban forests, large urban park, heritage garden), and two I-NBS (i.e., intensive green roof, semi-intensive green roof). Hence, thanks to these co-benefits, these seven NBS can be potentially implemented throughout 94% of the Italian territory, thus ensuring good performances employing less than one-third of the available NBS. On the contrary, among P-NBS, urban orchards and planter green walls show the lowest PS in all groups, thus representing a sub-optimal solution for addressing the three environmental challenges considered here (Table 2).

Figure 4 shows a specific focus on the distribution of impervious land cover within the spatial groups. Among the 20,000 km^2 of built-up areas in Italy, 8100 km^2 are occupied by the AIR-CLIM group (42.8%), 4700 km^2 by the AIR group (24.8%), over 2200 km^2 by the ALL and CLIM groups (12.7 and 11.7%, respectively), about 750 km^2 by the NoChal group (4%), and less than 450 km^2 by the AIR-WAT, CLIM-WAT, and WAT groups. Therefore, combining these coverages with the NBS having high-PS in each spatial group, we obtained the overall area where both P-NBS and I-NBS could potentially be implemented to address multiple challenges. With specific regard to the I-NBS: intensive and semi-intensive green roof (18,309 km^2), street trees (17,819 km^2), green façade (17,566 km^2), green wall system and vertical mobile garden (15,137 km^2), private gardens (13,127 km^2), and extensive green roof (13,085 km^2) (Figure 4).

Table 2. Twenty-four Nature-Based Solutions (NBS) classified as implementable in impervious (I-NBS) and permeable (P-NSB) land covers. Performance Score (PS) ranges between 0 (low performance) and 1 (high performance) for each NBS in the 7 groups, namely: AIR, CLIM, WAT, AIR-WAT, AIR-CLIM, CLIM-WAT, ALL. The black triangles mark NBS with high-PS (PS > 0.66). The NoChal group is not included considering we assumed that no new NBS are needed.

Nature Based Solutions	Performance Score (PS)						
I-NBS	**AIR**	**CLIM**	**WAT**	**AIR-CLIM**	**AIR-WAT**	**CLIM-WAT**	**ALL**
Extensive green roofs	0.5	▲0.9	0.6	▲0.7	0.5	▲0.7	▲0.7
Green walls system	▲1.0	▲0.8	0.0	▲0.9	0.5	0.4	0.6
Green façades	▲1.0	▲1.0	0.2	▲1.0	0.6	0.6	▲0.7
Intensive green roofs	0.7	▲0.9	▲0.8	▲0.8	▲0.8	▲0.9	▲0.8
Planter green walls	0.5	0.5	0.0	0.5	0.3	0.3	0.3
Pocket gardens/parks	0.6	0.6	▲0.8	0.6	▲0.7	▲0.7	▲0.7
Private gardens	0.5	▲1.0	▲0.8	▲0.8	0.6	▲0.9	▲0.8
Raingardens	0.4	0.3	▲0.8	0.4	0.6	0.6	0.5
Semi-intensive green roofs	▲0.7	▲0.8	▲1.0	▲0.8	▲0.8	▲0.9	▲0.8
Street trees	▲0.8	▲0.9	0.4	▲0.8	0.6	▲0.7	▲0.7
Swales	0.6	0.2	▲0.9	0.4	▲0.7	0.5	0.6
Vegetated grid paves	0.2	0.5	▲0.8	0.3	0.5	0.6	0.5
Vegetated pergola	0.5	▲0.8	0.3	0.6	0.4	0.5	0.5
Vertical mobile garden	▲1.0	▲0.9	0.0	▲1.0	0.5	0.5	0.6
P-NBS							
(Wet)Retention Ponds	▲0.8	0.6	▲1.0	▲0.7	▲0.9	▲0.8	▲0.8
Community gardens	0.3	0.5	▲0.8	0.4	0.6	▲0.7	0.6
Constructed wetlands	0.0	0.3	▲1.0	0.1	0.5	0.6	0.4
Green Corridors	▲1.0	▲1.0	▲0.7	▲1.0	▲0.8	▲0.8	▲0.9
Heritage gardens	▲1.0	▲1.0	▲1.0	▲1.0	▲1.0	▲1.0	▲1.0
Infiltration basins	▲0.8	▲0.8	▲1.0	▲0.8	▲0.9	▲0.9	▲0.9
Large urban parks	▲1.0	▲1.0	▲0.9	▲1.0	▲1.0	▲1.0	▲1.0
Shelters for biodiversity	▲0.8	0.0	▲0.7	0.4	▲0.7	0.3	0.5
Urban forests	▲1.0	▲0.9	▲0.8	▲0.9	▲0.9	▲0.9	▲0.9
Urban orchards	0.3	0.2	0.5	0.3	0.4	0.3	0.3

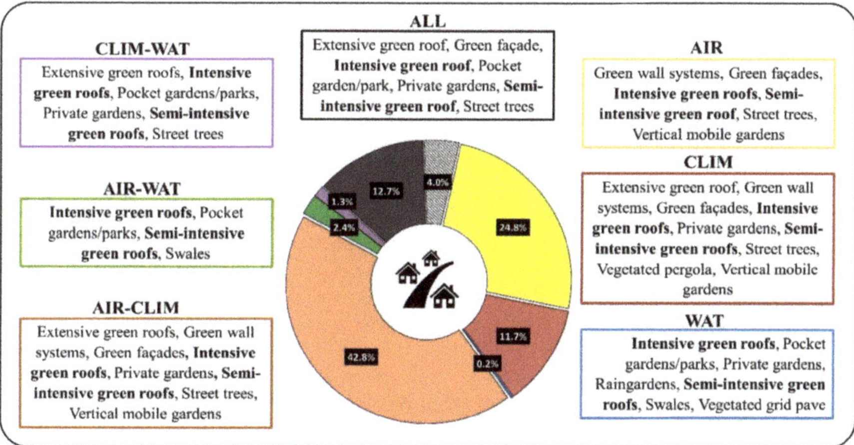

Figure 4. The pie chart shows the distribution of the spatial groups within the total impervious surface in Italy; the striped pattern represents the NoChal group. The I-NBS with high PS are reported in correspondence of each spatial group, with the I- NBS showing high PS simultaneously in all spatial groups marked in bold.

5. Discussion

Our framework has implications for the future development of cross-scale strategies to reach multiple national targets through NBS. It highlights the need to consider the multiple challenges to tackle as a key criterion to improve the NBS co-benefits and cost-effectiveness. Current NBS (or related concept) planning frameworks usually tend to focus on a single ES supply or address a specific challenge [41,68] and with a specific focus on the municipality scale [69–71]. However, our results highlighted that about 42% of the national territory shows multiple challenges simultaneously, and 50% of the population is exposed to these critical conditions. In these spatial groups (AIR-CLIM, AIR-WAT, CLIM-WAT, and ALL), multiple targets need to be achieved, through implementing NBS with the best performance to provide multiple ES. Consequently, the widespread distribution of areas under multiple challenges underlines that (i) in the political agenda, actions for air quality improvement might be coupled with those of climate change mitigation and adaptation [15]; (ii) considering both multiple ES demand (i.e., challenges to address) and multiple ES supply (i.e., NBS performance) has a key role to maximize the cost-effectiveness of interventions and the optimal use of the available space [8]. Europe—and the Member States—need to effectively leverage investments in NBS provided by the Green Deal, developing strategies to generate gains for adaptation, mitigation, disaster risk reduction, biodiversity, and health [37]. Therefore, the definition of a national intervention priority based on the intensity of challenges and population exposed [42] combined with the NBS performance ranking provided by our framework could help in this path, optimizing investment allocation from the national to the local governments.

Of the 24 NBS assessed, all spatial groups show from 11 to 16 NBS with high PS, both on impervious and permeable land covers. Providing a defined set of NBS is crucial for decision-makers and planners given the variety of NBS available [72], with different nomenclature, as well as the numerous barriers that may arise in urban areas from the planning stage to the site-scale design and implementation. The 24 NBS we considered in this work were selected from Castellar et al. [30], where, through using different workshops and surveys, they evaluated their performance to meet ten challenges, including the supply of all categories of ES. In the present work, some NBS may show similarities or overlapping results, being limited to only regulation ES (i.e., mitigation of air pollutants, heatwaves, and flood hazards). This could stand as a limitation; however, we decided not to further manipulate the nomenclature, thus leaving the possibility to replicate our methodology by also including other ES (e.g., provisioning, cultural, supporting) and other challenges (e.g., social cohesion).

Furthermore, the surfaces we evaluated as potentially available for the implementation of high-performing NBS do not necessarily correspond with the real space availability. Due to the broad scale and the aim of the work, we did not consider, e.g., archaeological constraints, protected areas, limited space in historical centers, that could decrease the suitability and space availability for NBS implementation. Therefore, for the local-scale implementation of the NBS, an in-depth assessment is necessary to include many other biophysical, economic, and social variables. To conduct a more detailed analysis and support the local scale governance to overcome the over-mentioned barriers, other layers could be useful, e.g., implementation and maintenance costs, the urban form, endemic vegetation, the public opinion, and many others that would be out of the scope and scale of this study.

However, according to the aim of this work, the incidence of land covers (i.e., impervious and permeable) in each spatial group, combined with the population density, suggests (i) which combination of factors is most related to the built-up areas, (ii) which risks the population is mainly exposed to, (iii) where to localize the interventions, and (iv) the number of beneficiaries of the expected increase in ES supply. On the one hand, we found spatial groups showing both high incidences of impervious surfaces and high population density, where I-NBS might be preferred. On the other hand, we found spatial groups with a lower incidence of permeable surfaces and low population density, where widespread and large-scale P-NBS (e.g., large urban parks, urban forests) in the territorial matrix could be more adequate. For example, impervious surfaces in the CLIM-WAT group have a built-up area's incidence eight times higher than in the AIR group (24% vs. 3%) and a population density twelve times greater too. This result suggests that investing equal resources (e.g., budget) in the first group, I-NBS, would affect more beneficiaries in a smaller area, mitigating both heatwaves and flood hazards, hence addressing two challenges simultaneously. These findings are particularly helpful since limited available space can act as barriers to NBS implementation, especially in urban areas where land is a scarce and expensive resource [21]. Potentially, some I-NBS (i.e., vertical green), even if less performing than others in P-NBS, have the advantage to occupy space often unemployed [73] and consequently not contributing to exacerbating conflicts around open space (e.g., land use change) in densely built-up areas [74].

5.1. Nature Based Solutions Implementable in Impervious Land Cover

Our results show that intensive and semi-intensive green roofs can potentially be implemented on 18,300 km^2, showing high PS in all spatial groups and hence standing as the most versatile and effective NBS among all the I-NBS assessed in this work. Although intensive and semi-intensive green roofs were initially designed for water management [75], due to their more deeply planted vegetation [73], they are also proved to positively contribute to climate mitigation, air quality, and biodiversity.

In terms of coverage and performance, among I-NBS, street trees and private gardens have high PS in five spatial groups and can be implemented, respectively, across 17,800 km^2 and 13,100 km^2, usually close to buildings, houses, and streets. Street trees show the best performance for the mitigation of air pollutants (AIR) and heatwaves (CLIM), both individually and combined (AIR-CLIM). Furthermore, the species selection can help both to mitigate pollutants [76,77] and to provide shade by the crown coverage [78,79]. Otherwise, when heatwaves and flood hazards need to be simultaneously mitigated, private gardens are more effective than street trees, contributing both to water management through the broadest unsealed soils, and to enhance air circulation and cooling through plant transpiration and shading [80]. Similarly, vertical green solutions (i.e., green façades, vertical mobile gardens, and green wall systems), are mainly reliable to stock air pollutants [81] and heatwave mitigation [82,83]. These I-NBS perform better in AIR, CLIM, and AIR-CLIM groups, covering potentially 17,500 km^2 of impervious surface. On the other hand, extensive green roofs show high PS for heatwaves and flood hazards [84], hence represented within AIR-CLIM, CLIM-WAT, CLIM, and ALL groups, and covering 13,000 km^2 of impervious surface.

5.2. Nature Based Solutions Implementable in Permeable Land Cover

Unlike the built-up areas where NBS Type 3 are usually considered, before implementing P-NBS, it is first necessary to evaluate the current land uses to consider their conservation (Type 1) and management (Type 2) instead of the substitution with new ecosystems (Type 3). This is following what was observed by Sarabi et al. [21], i.e., the entity of interventions required in NBS increases when moving (closer) to the center of built-up contexts, and vice versa. Therefore, in the case of currently forested areas, the objective should focus on their preservation, restoration, or enhancement to maximize ES supply (Type 1 and Type 2 [67]). This is a relevant option, for example, in the case of the AIR group, mainly occupied by forested areas. This is in line with the trajectory identified by the EU Green Deal, where, along with the protection and management of existing forests, urban, peri-urban, and agricultural areas need to be integrated with additional trees (i.e., 3 billion trees [23]). Accordingly, the assessed P-NBS can be applied in marginal, abandoned, unproductive, peri-urban areas [85–87], since they are considered as new ecosystems (Type 3 [67]).

Despite the finding that five P-NBS have high PS in all spatial groups (green corridors, large urban parks, heritage gardens, infiltration basins, and urban forests), they are similar to each other for regulating ES supply and thus addressing the challenges we considered in this work. The choice to implement one P-NBS instead of others can be related to the supply of other ES categories (cultural, provisioning, and support) as well as to other policy and planning issues (e.g., people perceptions, recreation needs) and budgetary constraints. Particularly, large urban parks and heritage gardens refer to large green areas (>0.5 ha) with mixed land uses (i.e., forests, grasslands, ponds). The first ones are mainly oriented to provide a variety of recreational facilities, mainly addressing the social demands of the residents, while the second ones aim to preserve outstanding historical, cultural, aesthetic, or scientific value [30]. Moreover, co-benefits and multifunctionality (i.e., multiple ES supply) of the single NBS can be enhanced by adding some improvements, usually including tree planting. As an example, infiltration basins can be partially forested to fulfill other functions such as providing recreational spaces for inhabitants, increasing biodiversity and ecological connectivity [88]. Similarly, green corridors are usually renatured areas of derelict infrastructure (i.e., railway) or placed along rivers. They can be afforested where there is the need to enhance landscape connectivity and ecological restoration [70,89,90]. Furthermore, their social role can be emphasized by ameliorating the availability and accessibility of currently vacant and underused land in urban contexts [91]. Therefore, the five P-NBS considered here already include—or could include—individual trees and/or groups of trees, as they are considered to be the best natural elements to increase the spectrum of ES provided [26,79,85,92–94].

6. Conclusions

The environmental challenges addressed in this work can adversely affect human health and well-being, with associated mitigation costs. Accordingly, our work contains a novel framework that will help both the national government and the municipalities to identify NBS able to maximize the ES supply while addressing multiple challenges. Analogously to the already proposed "National Strategy on Urban Greenspaces" [59], this work can provide a strategic vision at the national scale, but it can be consulted and adopted by all municipalities as a common roadmap, also helpful in facing the recurring problematic of planning silos. Indeed, the relevance of our framework is not just focused on the NBS application at the local scale but also shows a great impact on a wider scale (e.g., national and regional).

On a national scale, the framework proposed here can reliably (i) identify the areas showing a simultaneous demand for the achievement of multiple national targets; (ii) spatially orient the new investment needed to mitigate the challenges (e.g., EU Green Deal); and (iii) support the NBS selection that provides more co-benefits, playing a crucial role in increasing budget allocations efficiency.

On a municipal scale, the NBS ranking can be used as a guideline for further specific planning and design activities based on local issues, barriers, and peculiarities, while remaining consistent with national targets.

Italy is currently allocating funds in the 14 metropolitan cities to implement urban forests. Our results confirm that urban forests are among the best performing NBS, and Di Pirro et al. [42] argue that reforestation programs could also be expanded to other municipalities with few additional resources (+7.5% of the national territory) but involving an extra 46% of the national population. Although trees and forests (especially urban ones) are considered by many authors as the best solution to address environmental challenges [79,85,92,93], our work also proposes a list of performing I-NBS (e.g., green roofs) that can be implemented on sealed surfaces. These can help mitigate environmental stressors by using impervious surfaces i) that are usually unemployed (e.g., gravel or bitumen roofs) and ii) that could even exacerbate the challenges due to their physical characteristics (e.g., thermal emittance, reduced infiltration capacity) [95]. This option can also contribute to mitigating the negative effects related to soil sealing, which is a remarkable issue in the EU [96,97], enhancing the values of interstitial and leftover spaces [87]. However, the technical feasibility and costs related to these I-NBS and their widespread implementation must be evaluated according to the specific local conditions [73]. Finally, at the local scale, additional co-benefits (i.e., energy savings, biodiversity increase, and social cohesion), as well as possible disservices (i.e., BVOC emissions, decrease in wind velocity, gentrification), should also be included for a more overarching assessment [94,98,99]. Planning and managing NBS can be approached holistically [40], considering diverse benefits concerning different spatial–temporal scales. The multi-scale approach can help in considering different stakeholders as well as social, economic, and biophysical characteristics that matter in the benefit provision and are thus better included in decision-making related to national, regional, city/site-scale spatial plans [100].

Author Contributions: Conceptualization, E.D.P. and L.S.; methodology, E.D.P., L.S. and J.A.C.C.; software, formal analysis and data curation, E.D.P.; validation, E.D.P., L.S., J.A.C.C., G.S. and B.L.; investigation, E.D.P., L.S. and G.S.; resources, M.M. and B.L.; writing—original draft preparation, E.D.P.; writing—review and editing, E.D.P., L.S., J.A.C.C., G.S., M.M. and B.L.; visualization, E.D.P., G.S. and M.M.; supervision, M.M. and B.L.; project administration and funding acquisition, B.L. All authors have read and agreed to the published version of the manuscript.

Funding: This work has received funding from the research project "Establishing Urban FORest based solutions In Changing Cities" (EUFORICC), cod 20173RRN2S, funded by the PRIN 2017 program of the Italian Ministry of University and Research (project coordinator: C. Calfapietra).

Institutional Review Board Statement: Not applicable.

Informed Consent Statement: Not applicable.

Data Availability Statement: The data presented in this study are available on request from the corresponding author.

Conflicts of Interest: The authors declare no conflict of interest.

References

1. Raymond, C.M.; Frantzeskaki, N.; Kabisch, N.; Berry, P.; Breil, M.; Nita, M.R.; Geneletti, D.; Calfapietra, C. A framework for assessing and implementing the co-benefits of nature-based solutions in urban areas. *Environ. Sci. Policy* **2017**, *77*, 15–24. [CrossRef]
2. European Commission. *Towards an EU Research and Innovation Policy Agenda for Nature-Based Solutions & Re-Naturing Cities. Final Report of the Horizon 2020 Expert Group on "Nature-Based Solutions and Re-Naturing Cities"*; Publications Office of the European Union: Luxembourg, 2015; ISBN 978-92-79-46051-7.
3. Snep, R.P.H.; Voeten, J.G.W.F.; Mol, G.; Van Hattum, T. Nature Based Solutions for Urban Resilience: A Distinction between No-Tech, Low-Tech and High-Tech Solutions. *Front. Environ. Sci.* **2020**, *8*, 259. [CrossRef]
4. Dawson, R.; Wyckmans, A.; Heidrich, O.; Köhler, J.; Dobson, S.; Feliu, E. *Understanding Cities: Advances in Integrated Assessment of Urban Sustainability*; Centre for Earth Systems Engineering Research (CESER), Newcastle University: Newcastle, UK, 2014; ISBN 9780992843700.
5. European Union. EU Law and Related Documents. Available online: http://eur-lex.europa.eu (accessed on 12 January 2022).
6. Dumitru, A.; Wendling, L. *Evaluating the Impact of Nature-Based Solutions: A Handbook for Practitioners*; European Commission (EC): Brussels, Belgium, 2021; ISBN 9789276229612.
7. Khomenko, S.; Cirach, M.; Pereira-Barboza, E.; Mueller, N.; Barrera-Gómez, J.; Rojas-Rueda, D.; de Hoogh, K.; Hoek, G.; Nieuwenhuijsen, M. Premature mortality due to air pollution in European cities: A health impact assessment. *Lancet Planet. Health* **2021**, *5*, e121–e134. [CrossRef]
8. Wild, T.C.; Henneberry, J.; Gill, L. Comprehending the multiple 'values' of green infrastructure—Valuing nature-based solutions for urban water management from multiple perspectives. *Environ. Res.* **2017**, *158*, 179–187. [CrossRef] [PubMed]
9. Balzan, M.V.; Tomaskinova, J.; Collier, M.; Dicks, L.; Geneletti, D.; Grace, M.; Longato, D.; Sadula, R.; Stoev, P.; Sapundzhieva, A. Building capacity for mainstreaming nature-based solutions into environmental policy and landscape planning. *Res. Ideas Outcomes* **2020**, *6*, e58970. [CrossRef]
10. Kabisch, N.; Korn, H.; Stadler, J.; Bonn, A. *Nature-Based Solutions to Climate Change Adaptation in Urban Areas*; Kabisch, N., Korn, H., Stadler, J., Bonn, A., Eds.; Theory and Practice of Urban Sustainability Transitions; Springer International Publishing: Cham, Switzerland, 2017; ISBN 978-3-319-53750-4.
11. Kabisch, N.; Frantzeskaki, N.; Pauleit, S.; Naumann, S.; Davis, M.; Artmann, M.; Haase, D.; Knapp, S.; Korn, H.; Stadler, J.; et al. Nature-based solutions to climate change mitigation and adaptation in urban areas: Perspectives on indicators, knowledge gaps, barriers, and opportunities for action. *Ecol. Soc.* **2016**, *21*, 39. [CrossRef]

12. Escobedo, F.J.; Giannico, V.; Jim, C.Y.; Sanesi, G.; Lafortezza, R. Urban forests, ecosystem services, green infrastructure and nature-based solutions: Nexus or evolving metaphors? *Urban For. Urban Green.* **2019**, *37*, 3–12. [CrossRef]
13. Cohen-Shacham, E.; Andrade, A.; Dalton, J.; Dudley, N.; Jones, M.; Kumar, C.; Maginnis, S.; Maynard, S.; Nelson, C.R.; Renaud, F.G.; et al. Core principles for successfully implementing and upscaling Nature-based Solutions. *Environ. Sci. Policy* **2019**, *98*, 20–29. [CrossRef]
14. Albert, C.; Schröter, B.; Haase, D.; Brillinger, M.; Henze, J.; Herrmann, S.; Gottwald, S.; Guerrero, P.; Nicolas, C.; Matzdorf, B. Addressing societal challenges through nature-based solutions: How can landscape planning and governance research contribute? *Landsc. Urban Plan.* **2019**, *182*, 12–21. [CrossRef]
15. Raymond, C.M.; Pam, B.; Breil, M.; Nita, M.R.; Kabisch, N.; de Bel, M.; Enzi, V.; Frantzeskaki, N.; Geneletti, D.; Cardinaletti, M.; et al. *An Impact Evaluation Framework to Support Planning and Evaluation of Nature-based Solutions Projects*; Centre for Ecology and Hydrology: Wallingford, UK, 2017; ISBN 9781906698621.
16. Faivre, N.; Fritz, M.; Freitas, T.; de Boissezon, B.; Vandewoestijne, S. Nature-Based Solutions in the EU: Innovating with nature to address social, economic and environmental challenges. *Environ. Res.* **2017**, *159*, 509–518. [CrossRef]
17. Kooijman, E.D.; McQuaid, S.; Rhodes, M.L.; Collier, M.J.; Pilla, F. Innovating with nature: From nature-based solutions to nature-based enterprises. *Sustainability* **2021**, *13*, 1263. [CrossRef]
18. Frantzeskaki, N.; McPhearson, T.; Collier, M.J.; Kendal, D.; Bulkeley, H.; Dumitru, A.; Walsh, C.; Noble, K.; Van Wyk, E.; Ordóñez, C.; et al. Nature-based solutions for urban climate change adaptation: Linking science, policy, and practice communities for evidence-based decision-making. *Bioscience* **2019**, *69*, 455–466. [CrossRef]
19. The World Bank. *Implementing Nature Based Flood Protection*; The World Bank: Washington, DC, USA, 2017.
20. Mendonça, R.; Roebeling, P.; Fidélis, T.; Saraiva, M. Policy instruments to encourage the adoption of nature-based solutions in urban landscapes. *Resources* **2021**, *10*, 81. [CrossRef]
21. Sarabi, S.E.; Han, Q.; Romme, A.G.L.; de Vries, B.; Wendling, L. Key enablers of and barriers to the uptake and implementation of nature-based solutions in urban settings: A review. *Resources* **2019**, *8*, 121. [CrossRef]
22. Chausson, A.; Turner, B.; Seddon, D.; Chabaneix, N.; Girardin, C.A.J.; Kapos, V.; Key, I.; Roe, D.; Smith, A.; Woroniecki, S.; et al. Mapping the effectiveness of nature-based solutions for climate change adaptation. *Glob. Chang. Biol.* **2020**, *26*, 6134–6155. [CrossRef] [PubMed]
23. European Commission. *The 3 Billion Tree Planting Pledge for 2030*; European Commission: Brussels, Belgium, 2021.
24. Seymour, F. Seeing the Forests as well as the (Trillion) Trees in Corporate Climate Strategies. *One Earth* **2020**, *2*, 390–393. [CrossRef]
25. Goffner, D.; Sinare, H.; Gordon, L.J. Correction to: The great Green Wall for the Sahara and the Sahel initiative as an opportunity to enhance resilience in Sahelian landscapes and livelihoods. *Reg. Environ. Chang.* **2019**, *19*, 2139–2140. [CrossRef]
26. McDonald, R.; Aljabar, L.; Aubuchon, C.; Birnbaum, H.; Chandler, C.; Toomey, B.; Daley, J.; Jimenez, W.; Trieschman, E.; Paque, J.; et al. *Funding Trees for Health: An Analysis of Finance and Policy Actions to Enable Tree Planting for Public Health*; Nature Conservancy: Arlington, VA, USA, 2017.
27. Stagakis, S.; Somarakis, G.; Chrysoulakis, N. *ThinkNature Nature Based Solutions Handbook*; ThinkNature Project Funded by EU Horizon 2020 Research and Innovation Program; European Union: Brussels, Belgium, 2019; 226p.
28. Collier, M.J.; Connop, S.; Foley, K.; Nedović-Budić, Z.; Newport, D.; Corcoran, A.; Crowe, P.; Dunne, L.; de Moel, H.; Kampelmann, S.; et al. Urban transformation with TURAS open innovations; opportunities for transitioning through transdisciplinarity. *Curr. Opin. Environ. Sustain.* **2016**, *22*, 57–62. [CrossRef]
29. ThinkNature. *Nature Based Solutions—Technical Handbook*; Part II; European Union: Brussels, Belgium, 2019.
30. Castellar, J.A.C.; Popartan, L.A.; Pueyo-Ros, J.; Atanasova, N.; Langergraber, G.; Säumel, I.; Corominas, L.; Comas, J.; Acuña, V. Nature-based solutions in the urban context: Terminology, classification and scoring for urban challenges and ecosystem services. *Sci. Total Environ.* **2021**, *779*, 146237. [CrossRef] [PubMed]
31. Wild, T.; Bulkeley, H.; Naumann, S.; Vojinovic, Z.; Calfapietra, C.; Whiteoak, K. *Nature-Based Solutions: State of the Art in EU-Funded Projects*; Publications Office of the European Union: Luxembourg, 2020; ISBN 978-92-76-17334-2.
32. Davis, M.; Abhold, K.; Mederake, L.; Knoblauch, D. NBS in European and National Policy Framework. *Naturvation* **2018**, *50*, 1–52.
33. UNEP. *Smart, Sustainable and Resilient Cities: The Power of Nature-based Solutions*; UNEP: Nairobi, Kenya, 2021; pp. 1–32.
34. IUCN. *Global Standard for Nature-Based Solutions: A User-Friendly Framework for the Verification, Design and Scaling Up of NbS: First edition*; International Union for Conservation of Nature (IUCN): Gland, Switzerland, 2020.
35. Nesshöver, C.; Assmuth, T.; Irvine, K.N.; Rusch, G.M.; Waylen, K.A.; Delbaere, B.; Haase, D.; Jones-Walters, L.; Keune, H.; Kovacs, E.; et al. The science, policy and practice of nature-based solutions: An interdisciplinary perspective. *Sci. Total Environ.* **2017**, *579*, 1215–1227. [CrossRef] [PubMed]
36. Albert, C.; Fürst, C.; Ring, I.; Sandström, C. Research note: Spatial planning in Europe and Central Asia—Enhancing the consideration of biodiversity and ecosystem services. *Landsc. Urban Plan.* **2020**, *196*, 103741. [CrossRef]
37. Veerkamp, C.; Ramieri, E.; Romanovska, L.; Zandersen, M.; Förster, J.; Rogger, M.; Martinsen, L. *Assessment Frameworks of Nature-Based Solutions for Climate Change Adaptation and Disaster Risk Reduction*; European Topic Centre on Climate Change Impacts, Vulnerability and Adaptation (ETC/CCA): Wageningen, The Netherlands, 2021.
38. Grace, M.; Balzan, M.; Collier, M.; Geneletti, D.; Tomaskinova, J.; Abela, R.; Borg, D.; Buhagiar, G.; Camilleri, L.; Cardona, M.; et al. Priority knowledge needs for implementing nature-based solutions in the Mediterranean islands. *Environ. Sci. Policy* **2021**, *116*, 56–68. [CrossRef]

39. Cramer, W.; Guiot, J.; Fader, M.; Garrabou, J.; Gattuso, J.-P.; Iglesias, A.; Lange, M.; Lionello, P.; Lla-sat, M.; Paz, S.; et al. Risks associated to climate and environmental changes in the Mediterranean region. *Br. J. Psychiatry* **2019**, *111*, 1009–1010.
40. Meerow, S. The politics of multifunctional green infrastructure planning in New York City. *Cities* **2020**, *100*, 102621. [CrossRef]
41. Meerow, S.; Newell, J.P. Spatial planning for multifunctional green infrastructure: Growing resilience in Detroit. *Landsc. Urban Plan.* **2017**, *159*, 62–75. [CrossRef]
42. Di Pirro, E.; Sallustio, L.; Sgrigna, G.; Marchetti, M.; Lasserre, B. Strengthening the implementation of national policy agenda in urban areas to face multiple environmental stressors: Italy as a case study. *Environ. Sci. Policy* **2022**, *129*, 1–11. [CrossRef]
43. Fischer, E.M.; Schär, C. Consistent geographical patterns of changes in high-impact European heatwaves. *Nat. Geosci.* **2010**, *3*, 398–403. [CrossRef]
44. WMO. Summer of Extremes: Floods, Heat and Fire. Available online: https://public.wmo.int/en/media/news/summer-of-extremes-floods-heat-and-fire (accessed on 15 September 2021).
45. European Environment Agency (EEA). *Air Quality in Europe—2020 Report*; EEA: Copenhagen, Denmark, 2020; ISBN 978-92-9480-292-7.
46. Sicard, P.; Agathokleous, E.; De Marco, A.; Paoletti, E.; Calatayud, V. Urban population exposure to air pollution in Europe over the last decades. *Environ. Sci. Eur.* **2021**, *33*, 28. [CrossRef] [PubMed]
47. SNPA. *Consumo di Suolo, Dinamiche Territoriali e Servizi Ecosistemici*; SNPA: Roma, Italy, 2019; Volume 8.
48. Sallustio, L.; Munafò, M.; Riitano, N.; Lasserre, B.; Fattorini, L.; Marchetti, M. Integration of land use and land cover inventories for landscape management and planning in Italy. *Environ. Monit. Assess.* **2016**, *188*, 48. [CrossRef]
49. Romano, B.; Zullo, F.; Fiorini, L.; Ciabò, S.; Marucci, A. Sprinkling: An Approach to Describe Urbanization Dynamics in Italy. *Sustainability* **2017**, *9*, 97. [CrossRef]
50. Amato, F.; Maimone, B.; Martellozzo, F.; Nolè, G.; Murgante, B. The Effects of Urban Policies on the Development of Urban Areas. *Sustainability* **2016**, *8*, 297. [CrossRef]
51. Sallustio, L.; Quatrini, V.; Geneletti, D.; Corona, P.; Marchetti, M. Assessing land take by urban development and its impact on carbon storage: Findings from two case studies in Italy. *Environ. Impact Assess. Rev.* **2015**, *54*, 80–90. [CrossRef]
52. Di Pirro, E.; Sallustio, L.; Capotorti, G.; Marchetti, M.; Lasserre, B. A scenario-based approach to tackle trade-offs between biodiversity conservation and land use pressure in Central Italy. *Ecol. Modell.* **2021**, *448*, 109533. [CrossRef]
53. Sallustio, L.; De Toni, A.; Strollo, A.; Di Febbraro, M.; Gissi, E.; Casella, L.; Geneletti, D.; Munafò, M.; Vizzarri, M.; Marchetti, M. Assessing habitat quality in relation to the spatial distribution of protected areas in Italy. *J. Environ. Manag.* **2017**, *201*, 129–137. [CrossRef]
54. Capotorti, G.; Guida, D.; Siervo, V.; Smiraglia, D.; Blasi, C. Ecological classification of land and conservation of biodiversity at the national level: The case of Italy. *Biol. Conserv.* **2012**, *147*, 174–183. [CrossRef]
55. Marchetti, M.; Motta, R.; Salbitano, F.; Vacchiano, G. Planting trees in Italy for the health of the planet. Where, how and why. *For. Riv. Selvic. Ed. Ecol. For.* **2019**, *16*, 59–65. [CrossRef]
56. Decree on Climate, Decreto Clima—Decreto Legge 14 Ottobre 2019, n. 111 (GU Serie Generale n.241 del 14-10-2019). 2019. Available online: https://www.gazzettaufficiale.it/eli/id/2019/10/14/19G00125/sg (accessed on 10 June 2021).
57. Air Quality Directive, Directive 2008/50/EC of The European Parliament and of the Council of 21 May 2008 on Ambient Air Quality and Cleaner Air for Europe. Available online: http://eur-lex.europa.eu/legal-content/en/ALL/?uri=CELEX:32008L0050 (accessed on 10 January 2021).
58. Salbitano, F.; Sanesi, G. Decree on Climate 2019. What resources support forests and silviculture in our cities? *For. Riv. Selvic. Ed. Ecol. For.* **2019**, *16*, 74–76. [CrossRef]
59. MATTM. *Strategia Nazionale del Verde Urbano*; Ministero Della Transizione Ecologica: Roma, Italy, 2018.
60. Sallustio, L.; Lasserre, B.; Blasi, C.; Marchetti, M. Infrastrutture verdi contro il consumo di suolo. *Reticula* **2020**, *25*, 21–31.
61. Di Napoli, C.; Pappenberger, F.; Cloke, H.L. Verification of heat stress thresholds for a health-based heat-wave definition. *J. Appl. Meteorol. Climatol.* **2019**, *58*, 1177–1194. [CrossRef]
62. Baró, F.; Haase, D.; Gómez-Baggethun, E.; Frantzeskaki, N. Mismatches between ecosystem services supply and demand in urban areas: A quantitative assessment in five European cities. *Ecol. Indic.* **2015**, *55*, 146–158. [CrossRef]
63. Trigila, A.; Iadanza, C.; Bussettini, M.; Lastoria, B. *Rapporto sul Dissesto Idrogeologico in Italia*; ISPRA: Rome, Italy, 2018.
64. Horálek, J.; Schreiberová, M.; Schneider, P.; Kurfürst, P.; Schovánková, J.; Ďoubalová, J. *European Air Quality Maps for 2017*; European Topic Centre on Air Pollution, Noise and Industrial Pollution: Bilthoven, The Netherlands, 2019.
65. European Environment Agency Copernicus Land Monitoring Service High Resolution Land Cover Characteristics. Imperviousness 2018, Imperviousness Change 2015–2018 and Built-Up 2018. 2018. Permalink: 7860bc42f4c1494599f1e135c832788c. Available online: https://www.eea.europa.eu/data-and-maps/data/copernicus-land-monitoring-service-eu-dem (accessed on 15 September 2021).
66. Congedo, L.; Sallustio, L.; Munafò, M.; Ottaviano, M.; Marchetti, M.; Congedo, L.; Sallustio, L.; Munafò, M.; Ottaviano, M. Copernicus high-resolution layers for land cover classification in Italy. *J. Maps* **2016**, *12*, 1195–1205. [CrossRef]
67. Eggermont, H.; Balian, E.; Azevedo, J.M.N.; Beumer, V.; Brodin, T.; Claudet, J.; Fady, B.; Grube, M.; Keune, H.; Lamarque, P.; et al. Nature-based solutions: New influence for environmental management and research in Europe. *Gaia* **2015**, *24*, 243–248. [CrossRef]
68. Kremer, P.; Hamstead, Z.; Haase, D.; McPhearson, T.; Frantzeskaki, N.; Andersson, E.; Kabisch, N.; Larondelle, N.; Rall, E.L.; Voigt, A.; et al. Key insights for the future of urban ecosystem services research. *Ecol. Soc.* **2016**, *21*, 2. [CrossRef]

69. Marando, F.; Salvatori, E.; Sebastiani, A.; Fusaro, L.; Manes, F. Regulating Ecosystem Services and Green Infrastructure: Assessment of Urban Heat Island effect mitigation in the municipality of Rome, Italy. *Ecol. Modell.* **2019**, *392*, 92–102. [CrossRef]
70. Capotorti, G.; Del Vico, E.; Anzellotti, I.; Celesti-Grapow, L. Combining the Conservation of Biodiversity with the Provision of Ecosystem Services in Urban Green Infrastructure Planning: Critical Features Arising from a Case Study in the Metropolitan Area of Rome. *Sustainability* **2016**, *9*, 10. [CrossRef]
71. Manes, F.; Marando, F.; Capotorti, G.; Blasi, C.; Salvatori, E.; Fusaro, L.; Ciancarella, L.; Mircea, M.; Marchetti, M.; Chirici, G.; et al. Regulating Ecosystem Services of forests in ten Italian Metropolitan Cities: Air quality improvement by PM10 and O3 removal. *Ecol. Indic.* **2016**, *67*, 425–440. [CrossRef]
72. Croeser, T.; Garrard, G.; Sharma, R.; Ossola, A.; Bekessy, S. Choosing the right nature-based solutions to meet diverse urban challenges. *Urban For. Urban Green.* **2021**, *65*, 127337. [CrossRef]
73. Stovin, V. The potential of green roofs to manage urban stormwater. *Water Environ. J.* **2010**, *24*, 192–199. [CrossRef]
74. Zölch, T.; Henze, L.; Keilholz, P.; Pauleit, S. Regulating urban surface runoff through nature-based solutions—An assessment at the micro-scale. *Environ. Res.* **2017**, *157*, 135–144. [CrossRef] [PubMed]
75. Carter, T.; Jackson, C.R. Vegetated roofs for stormwater management at multiple spatial scales. *Landsc. Urban Plan.* **2007**, *80*, 84–94. [CrossRef]
76. Tallis, M.; Taylor, G.; Sinnett, D.; Freer-Smith, P. Estimating the removal of atmospheric particulate pollution by the urban tree canopy of London, under current and future environments. *Landsc. Urban Plan.* **2011**, *103*, 129–138. [CrossRef]
77. Sgrigna, G.; Baldacchini, C.; Dreveck, S.; Cheng, Z.; Calfapietra, C. Relationships between air particulate matter capture efficiency and leaf traits in twelve tree species from an Italian urban-industrial environment. *Sci. Total Environ.* **2020**, *718*, 137310. [CrossRef] [PubMed]
78. Armson, D.; Stringer, P.; Ennos, A.R. The effect of tree shade and grass on surface and globe temperatures in an urban area. *Urban For. Urban Green.* **2012**, *11*, 245–255. [CrossRef]
79. Livesley, S.J.; McPherson, E.G.; Calfapietra, C. The Urban Forest and Ecosystem Services: Impacts on Urban Water, Heat, and Pollution Cycles at the Tree, Street, and City Scale. *J. Environ. Qual.* **2016**, *45*, 119–124. [CrossRef] [PubMed]
80. Cabral, I.; Costa, S.; Weiland, U.; Bonn, A. *Etta Urban Gardens as Multifunctional Nature-Based Solutions for Societal Goals in a Changing Climate*; Kabisch, N., Korn, H., Stadler, J., Bonn, A., Eds.; Theory and Practice of Urban Sustainability Transitions; Springer International Publishing: Cham, Switzerland, 2017; ISBN 978-3-319-53750-4.
81. Ysebaert, T.; Koch, K.; Samson, R.; Denys, S. Green walls for mitigating urban particulate matter pollution—A review. *Urban For. Urban Green.* **2021**, *59*, 127014. [CrossRef]
82. Kántor, N.; Gál, C.V.; Gulyás, Á.; Unger, J. The Impact of Façade Orientation and Woody Vegetation on Summertime Heat Stress Patterns in a Central European Square: Comparison of Radiation Measurements and Simulations. *Adv. Meteorol.* **2018**, *2018*, 2650642. [CrossRef]
83. Blanco, I.; Schettini, E.; Vox, G. Effects of vertical green technology on building surface temperature. *Int. J. Des. Nat. Ecodyn.* **2018**, *13*, 384–394. [CrossRef]
84. Susca, T.; Gaffin, S.R.; Dell'Osso, G.R. Positive effects of vegetation: Urban heat island and green roofs. *Environ. Pollut.* **2011**, *159*, 2119–2126. [CrossRef]
85. Salbitano, F.; Borelli, S.; Conigliaro, M.; Chen, Y. *Guidelines on Urban and Peri-Urban Forestry*; FAO: Rome, Italy, 2016; Volume 178, ISBN 9789251094426.
86. Capotorti, G.; De Lazzari, V.; Ortí, M.A. Local Scale Prioritisation of Green Infrastructure for Enhancing Biodiversity in Peri-Urban Agroecosystems: A Multi-Step Process Applied in the Metropolitan City of Rome (Italy). *Sustainability* **2019**, *11*, 3322. [CrossRef]
87. Kim, M.; Rupprecht, C.D.D.; Furuya, K. Typology and Perception of Informal Green Space in Urban Interstices: A Case Study of Ichikawa City, Japan. *Int. Rev. Spat. Plan. Sustain. Dev.* **2020**, *8*, 4–20. [CrossRef]
88. NRWM. Infiltration Basins; 2014. Project Publications are Available. Available online: http://www.nwrm.eu (accessed on 15 January 2022).
89. Valeri, S.; Zavattero, L.; Capotorti, G. Ecological connectivity in agricultural green infrastructure: Suggested criteria for fine scale assessment and planning. *Land* **2021**, *10*, 807. [CrossRef]
90. Capotorti, G.; Alós Ortí, M.M.; Copiz, R.; Fusaro, L.; Mollo, B.; Salvatori, E.; Zavattero, L. Biodiversity and ecosystem services in urban green infrastructure planning: A case study from the metropolitan area of Rome (Italy). *Urban For. Urban Green.* **2019**, *37*, 87–96. [CrossRef]
91. Zhang, Z.; Meerow, S.; Newell, J.P.; Lindquist, M. Enhancing landscape connectivity through multifunctional green infrastructure corridor modeling and design. *Urban For. Urban Green.* **2019**, *38*, 305–317. [CrossRef]
92. Endreny, T.; Santagata, R.; Perna, A.; De Stefano, C.; Rallo, R.F.; Ulgiati, S. Implementing and managing urban forests: A much needed conservation strategy to increase ecosystem services and urban wellbeing. *Ecol. Modell.* **2017**, *360*, 328–335. [CrossRef]
93. Hale, J.D.; Pugh, T.A.M.; Sadler, J.P.; Boyko, C.T.; Brown, J.; Caputo, S.; Caserio, M.; Coles, R.; Farmani, R.; Hales, C.; et al. Delivering a multi-functional and resilient urban forest. *Sustainability* **2015**, *7*, 4600–4624. [CrossRef]
94. McDonald, A.G.; Bealey, W.J.; Fowler, D.; Dragosits, U.; Skiba, U.; Smith, R.I.; Donovan, R.G.; Brett, H.E.; Hewitt, C.N.; Nemitz, E. Quantifying the effect of urban tree planting on concentrations and depositions of PM10 in two UK conurbations. *Atmos. Environ.* **2007**, *41*, 8455–8467. [CrossRef]

95. Marvuglia, A.; Koppelaar, R.; Rugani, B. The effect of green roofs on the reduction of mortality due to heatwaves: Results from the application of a spatial microsimulation model to four European cities. *Ecol. Modell.* **2020**, *438*, 109351. [CrossRef]
96. Romano, B.; Zullo, F.; Fiorini, L.; Marucci, A. Molecular No Smart-Planning in Italy: 8000 Municipalities in Action throughout the Country. *Sustainability* **2019**, *11*, 6467. [CrossRef]
97. Fiorini, L.; Zullo, F.; Marucci, A.; Romano, B. Land take and landscape loss: Effect of uncontrolled urbanization in Southern Italy. *J. Urban Manag.* **2019**, *8*, 42–56. [CrossRef]
98. Bodnaruk, E.W.; Kroll, C.N.; Yang, Y.; Hirabayashi, S.; Nowak, D.J.; Endreny, T.A. Where to plant urban trees? A spatially explicit methodology to explore ecosystem service tradeoffs. *Landsc. Urban Plan.* **2016**, *157*, 457–467. [CrossRef]
99. Almenar, J.B.; Rugani, B.; Geneletti, D.; Brewer, T. Integration of ecosystem services into a conceptual spatial planning framework based on a landscape ecology perspective. *Landsc. Ecol.* **2018**, *33*, 2047–2059. [CrossRef]
100. Demuzere, M.; Orru, K.; Heidrich, O.; Olazabal, E.; Geneletti, D.; Orru, H.; Bhave, A.G.; Mittal, N.; Feliu, E.; Faehnle, M. Mitigating and adapting to climate change: Multi-functional and multi-scale assessment of green urban infrastructure. *J. Environ. Manag.* **2014**, *146*, 107–115. [CrossRef] [PubMed]

 *forests*

MDPI

Article

Significant Loss of Ecosystem Services by Environmental Changes in the Mediterranean Coastal Area

Adriano Conte [1], Ilaria Zappitelli [2], Lina Fusaro [1], Alessandro Alivernini [2], Valerio Moretti [2], Tiziano Sorgi [2], Fabio Recanatesi [3] and Silvano Fares [1,*]

1 National Research Council of Italy (CNR), Institute of Bioeconomy (IBE), 00185 Rome, Italy; adriano.conte@ibe.cnr.it (A.C.); lina.fusaro@cnr.it (L.F.)
2 Council for Agricultural Research and Economics (CREA), Research Centre for Forestry and Wood, 00166 Rome, Italy; ilaria.zappitelli@crea.gov.it (I.Z.); alessandro.alivernini@crea.gov.it (A.A.); valerio.moretti@crea.gov.it (V.M.); tiziano.sorgi@crea.gov.it (T.S.)
3 Department of Agricultural & Forestry Sciences (DAFNE), University of Tuscia, 01100 Viterbo, Italy; fabio.rec@unitus.it
* Correspondence: silvano.fares@cnr.it

Abstract: Mediterranean coastal areas are among the most threated forest ecosystems in the northern hemisphere due to concurrent biotic and abiotic stresses. These may affect plants functionality and, consequently, their capacity to provide ecosystem services. In this study, we integrated ground-level and satellite-level measurements to estimate the capacity of a 46.3 km^2 Estate to sequestrate air pollutants from the atmosphere, transported to the study site from the city of Rome. By means of a multi-layer canopy model, we also evaluated forest capacity to provide regulatory ecosystem services. Due to a significant loss in forest cover, estimated by satellite data as −6.8% between 2014 and 2020, we found that the carbon sink capacity decreased by 34% during the considered period. Furthermore, pollutant deposition on tree crowns has reduced by 39%, 46% and 35% for PM, NO_2 and O_3, respectively. Our results highlight the importance of developing an integrated approach combining ground measurements, modelling and satellite data to link air quality and plant functionality as key elements to improve the effectiveness of estimate of ecosystem services.

Keywords: ecosystem services; air pollution removal; carbon sequestration

Citation: Conte, A.; Zappitelli, I.; Fusaro, L.; Alivernini, A.; Moretti, V.; Sorgi, T.; Recanatesi, F.; Fares, S. Significant Loss of Ecosystem Services by Environmental Changes in the Mediterranean Coastal Area. *Forests* **2022**, *13*, 689. https://doi.org/10.3390/f13050689

Academic Editor: Pete Bettinger

Received: 28 March 2022
Accepted: 26 April 2022
Published: 28 April 2022

1. Introduction

The provisioning of ecosystem services (ES) by urban and peri-urban forests has direct effect on the social, economic and environmental benefits. Regulating ecosystem Services (RES) represent the benefits to mankind deriving from the regulation of ecosystem processes [1]. The role of forests as nature-based solutions to mitigate the effects of climate change has been largely discussed [2], becoming a fundamental part of the proposal formulated at international level during important summits as the G20 (Rome) and COP26 (Glasgow) that fix the ambitious goals of planting 1 trillion trees (Rome G20) and stop deforestation (COP26 Glasgow) by 2030.

Estimating the contribution of "process-driven" RES such as air quality improvement, water purification, and climate regulation to human well-being is typically made by models [3], often challenged by a lack of input data, especially for estimations at broad scales [4].

Canopy models help unravel the factors that influence the response of the forest ecosystems in different environmental conditions by testing scenarios [5] or evaluating the impact of different stressors [6]. Such models need climatic and ecophysiological inputs that are often unavailable, and direct measurements needs to be performed to obtain reliable estimates [7]. Meteorological and air quality data for big cities are often available at hourly to daily resolution [8–10], and when missing (rural areas), high-resolution gridded datasets can be used when proper spatial and temporal resolutions are available [11,12]. Better estimation of the biosphere–atmosphere interactions has been observed by multi-layer canopy models [13]. These models are more complex than single-layers (or big-leaf models) and often require species-specific ecophysiological parameters to properly estimate plant's capacity to provide RES [14]. Stomatal conductance is a key parameter not only for carbon assimilation but also to estimate the uptake of pollutants as tropospheric ozone (O_3), nitrogen dioxide (NO_2) and sulphur dioxide (SO_2) [15,16]. Therefore, stomatal regulation processes need to be estimated accurately. This is particularly relevant for the Mediterranean region, where dry deposition is the dominant pathway [17] and stomatal conductance is limited by stressors such as drought and ozone exposure [18,19]. On one hand, coupled photosynthesis–stomatal conductance (*A-gs*) models such the semi-empirical model proposed by Ball, Woodrow & Berry (BWB) [20], or the theoretical optimal stomatal behavior models proposed by Medlyn et al. [21] and by Katul et al. [22] are best indicated for simulating leaf-level gas exchanges [23]; on the other hand, these models require physiological parameters such as carboxylation velocity (Vc_{max}) and light use efficiency (J_{max}) not always available in literature. To overcome this issue, direct field measurements are often required [10,14]. In addition, such complex models need biometric parameters to characterize the trees' structure or the total leaf area, which is a key parameter to upscaling fluxes at the canopy level [10]. Structural features determination requires an extensive campaigns effort that are difficult to realize at metropolitan scale. However, plant-trait databases and satellite-derived vegetation indices are effective means to derive at large spatial scale those essential information [24,25]. Indeed, vegetation indicators of key structural parameters such as the leaf area index (LAI, $m^2\ m^{-2}$) and canopy cover (%) were available from 2014, thanks to the monitoring activity of the Proba-V satellite [24]. The provisioning of RES is not costless for forests since air pollution represents a threat for vegetation as well. For instance, ozone damage can cause up to 10% of carbon loss in Mediterranean forest ecosystems [26,27].

In this work, we first highlighted reduction and gaps in percent canopy cover between three years of study (2014, 2019 and 2020) in the Presidential Estate of Castelporziano (Metropolitan Area of Rome), a vulnerable natural forest area elected as test site. The three years of study are characterized by different pluviometric regimes, ranging from a well-known wet year [28–30] to increasingly drier years in 2019 and 2020. Concerning pollutants exposure, 2020 differed from previous years since the Sars-CoV-2 lockdown status [31–35] produced a drastic decrease in emissions from road traffic. Concerning biotic stress, probably to limit the expansion of the infestation by *Tomicus destruens* and *Toumayella parvicornis*, several thinning campaigns were carried out within the Estate between 2017 and 2019 [36,37]. Therefore, a comparison between 2019 and 2020 may reveal the impact of altered anthropogenic emissions on the ecosystems RES provisioning capacity, and a comparison between 2014 and 2019–2020 may reveal the impact of both pathogenic infestation and management practices to limit its expansion. In a second step, we estimated the sequestration of carbon and air pollutants by means of the Aggregated InteRpreTation of the eneRgy balance and water dynamics for Ecosystem Services assessment (AIRTREE) multi-layer canopy model [14] parameterized with leaf-level measurements, satellite-derived structural data and georeferenced maps of vegetation.

Our goal was to answer the following scientific questions: (1) How much changes in environmental conditions affect forests' health status? (2) Do such changes have an impact on RES provisioning? The obtained results can help disentangle the processes that may have caused over the last three decades significant and diffuse decline in forest health conditions and natural renewal capacity.

2. Materials and Methods

2.1. Study Area

The Presidential Estate of Castelporziano is a natural reserve of about 60 km^2 (of which 85% are natural forests), 25 km away from the south-southwest area of Rome on the

Tyrrhenian Sea coast. The Estate represents a hotspot for biodiversity in the Mediterranean area, which hosts more than 1000 plant species [38,39].

Oak forests and mixed deciduous broad-leaved woods predominate, occupying 23.6 km^2 (40% of the woods), a surface higher than the evergreen oaks, and are represented by holm oaks (4.3 km^2) and cork oaks (2.4 km^2), by pine forests (9.1 km^2) and by Mediterranean maquis favored by anthropic land use changes in past centuries [40]. The most significant aspect that makes Castelporziano a territory now unique in the Mediterranean Basin is given by the presence of a specific and therefore very high genetic biodiversity, with 5037 species recorded, representing an exceptional wealth of flora and fauna on an area of just 60 square kilometers [41], (Figure 1).

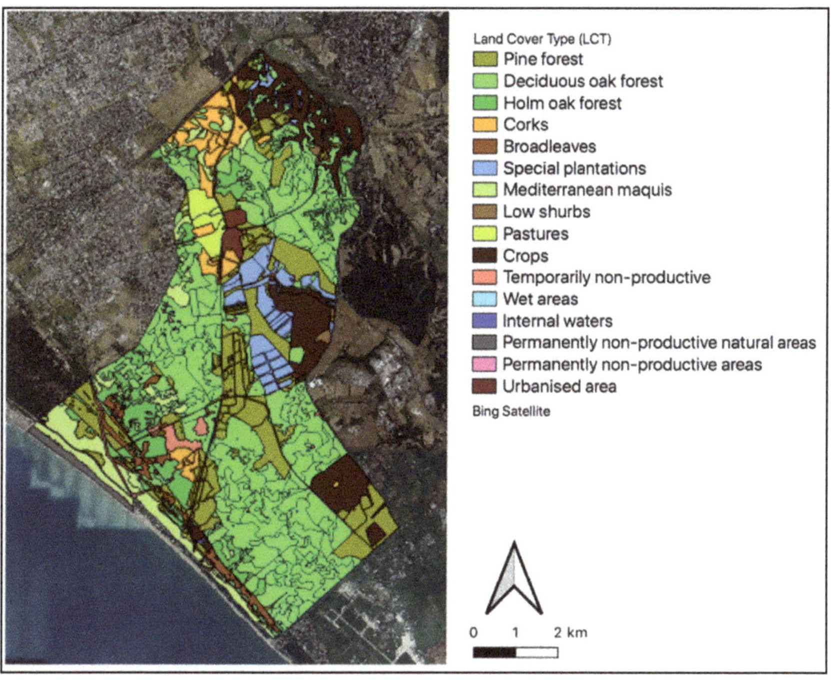

Figure 1. Vegetation map of the presidential Estate of Castelporziano, colored contours represent homogeneous vegetation, each color represent one Land Cover Type (LCT).

However, this valuable Green Infrastructure being located in the Metropolitan area of Rome is subjected to wide types of stressors, both natural and anthropogenic [42,43]. The intense urban sprawls nearby the Estate have caused an overexploitation of the water table and enhanced the seawater intrusion that exacerbate the natural drought that typically occurs during the summer. Moreover, high concentrations of pollutants that are transported over Castelporziano from the city center of Rome have been recorded. Indeed, the Estate receives plumes of air masses all day long from the sea and from the city of Rome because of the wind circulation, that follows a sea–land breeze regime, the dominant wind direction is S–SW during the morning and N–NE during the afternoon. The site can experience high summer levels of tropospheric ozone up to 90 ppb [44]. Soil presents a flat topography, and it can be divided in two main soil types (Figure S1): the coastal area, characterized by a sandy texture (77% sand, 14% silt, 9% clay) and low water-holding capacity, and the inland area, with a loamy-sandy soil type (47% sand, 29% silt, 24% clay), as shown in Figure S1 [45]. The study area is characterized by the typical Mediterranean climate, with pronounced seasonality. Summers are hot and dry, and winters are moderately cold.

Precipitations occur prevalently during winter, spring and autumn. Annual precipitation up to 1019 ± 105.5, 854 ± 130.8 and 507 ± 95.6 mm y^{-1} was observed in the 3 years of study in 2014, 2019 and 2020, respectively, considering averaged values for all the meteorological stations in the Estate. During the three years of study, mean annual temperature was of 16 °C with seasonal peaks in summer 2019 (Figure S2).

2.2. The AIRTREE Model

The Aggregated Interpretation of the Energy balance and water dynamics for Ecosystem services assessment (AIRTREE) model [14] is a one-dimensional multi-layer model that couples soil, plant and atmospheric processes to predict exchanges of CO_2, water, ozone and particulate matter (PM) between leaves and the atmosphere and integrates them through five layers to obtain fluxes at the canopy level. The model allows to estimate stomatal conductance (*gs*) and photosynthesis (*A*) through the Farquhar–Von Caemmerer–Berry model of photosynthesis (*FvCB* model) and the Ball, Woodrow and Berry stomatal conductance (BWB) model at different levels from the top to the bottom of the canopy [46]. The model accounts for oxidative limitations (i.e., drought and ozone stress) to gas exchanges (*A* and *gs*) through linear functions of soil water content [47] and stomatal ozone [48–50].

In this study NO_2, SO_2 and CO fluxes were calculated as:

$$F = V_d \cdot C \cdot 1800 \quad (1)$$

where F is the pollutant flux (µmol m^2 s^{-1}), V_d is the deposition velocity (m s^{-1}) and air pollutant concentration (µmol m^{-3}).

V_d was estimated through a resistance scheme [51,52]:

$$V_d = \frac{1}{R_a + R_b + R_c} \quad (2)$$

where R_a is the atmospheric resistance (see Fares et al. [14] for details), R_b is the boundary layer resistances (s m^{-1}) calculated according Pederson et al. [53] and R_c is the canopy resistance calculated according to Bidwell & Fraser [54] as:

$$R_b = 2(Sc)^{\frac{2}{3}} \cdot (Pr)^{-\frac{2}{3}} \cdot (ku)^{-1} \quad (3)$$

$$\frac{1}{R_c} = \frac{1}{r_s + r_m} + \frac{1}{r_{soil}} + \frac{1}{r_t} \quad (4)$$

where Sc is the Schmidt number, Pr is the Prandtl number (0.72), k is the von Karman constant (0.41), u is wind speed (m s^{-1}), r_s (s m^{-1}) is the inverse of g_s, the formulation of which is extensively described in Fares et al. [14]. Soil resistance r_{soil} (s m^{-1}) was set equal to 2941 and 2000 (m s^{-1}) during vegetative and non-vegetative periods, respectively. Mesophyll resistance r_m (s m^{-1}) was set to zero for SO_2 [55] and 600 for NO_2 [56]. Cuticular resistance r_t (s m^{-1}) was set to 8×10^3 and 2×10^4 for SO_2 and NO_2, respectively [56]. Rc for CO was set equal to 5×10^4 and 1×10^6 during vegetative and non-vegetative periods, respectively [54].

Concerning PM deposition, the model incorporates traditional atmospheric models to predict particle deposition extensively described in Fares et al. [10,14].

2.2.1. Acquisition of Meteorological Variables

For the entire Estate, we performed a Voronoi tessellation using data from a network of 6 meteorological stations where continuous monitoring of local climatic conditions have been carried out (Figure S3). Each station is equipped with sensors measuring air temperature and relative humidity, wind speed and direction, precipitation and solar radiation; for details on instrumentation, see Aromolo et al. [57]. In addition, the Estate hosts a flux tower for continuously monitoring gas exchanges (i.e., Carbon dioxide CO_2, ozone O_3, nitrogen dioxide NO_2, particulate matter PM) between the vegetation and the atmosphere in the coastal area, 1.5 km away from the Thyrrenean sea (see [30,44] for details). Just outside the Estate, within 1.5 km, there are 4 air quality monitoring stations maintained by the regional agency for the environmental protection (ARPA Lazio) that continuously monitor concentration of PM, NO_2, SO_2, CO, O_3 and other pollutants. Those air quality data were used as input for the AIRTREE model. The data were assumed to be exemplificative of Castelporziano site since a comparison between PM 2.5 concentration measured in 2014 and 2015 at the ICOS It-Cp2 flux site and average values measured at the 5 ARPA stations showed a good correlation (Pearson's $r > 0.7$, data not shown). The Voronoi tessellation method allowed us to generate a spatial-metric decomposition determined by distances to discrete set of elements in space (meteorological stations), and therefore, 6 Voronoi polygons were created for the whole Estate.

2.2.2. Acquisition of Vegetation Map

The vegetation map of the Castelporziano Estate (Figure 1) used in this study [58] is characterized by 7 groups of forested land cover type (LCT) (Other broadleaves; Low shrubs; Holm oak forests; Mediterranean maquis; Pine forests; Deciduous oak forests; Corks) and 9 non-forested LCT (Internal waters; Permanently non-productive areas; Permanently non-productive natural areas). To associate species-specific ecophysiological traits to each of the forested LCT group, we used the state-of-the-art of literature about the vegetation in Castelporziano to identify the most representative species included in each group [38]. Therefore, the values of carbon, PM and other pollutants deposition for each pixel (associated to a specific LCT) are the results of a single model simulation. For each model run, the AIRTREE model was parametrized for an ideal tree-type by synthetizing the ecophysiological characteristics of the dominant trees representative of the LCT (Table S1).

2.2.3. Retrieval of Biometric Vegetation Data

A raster map of the species associations was superimposed and aligned with the maps of Leaf Area Index (LAI), canopy cover and heights (Figure S4). Concerning LAI and canopy cover, we used 300 m resolution satellite images from Copernicus Global Land Service, the Earth Observation programme of the European Commission [59]. For each year of study (i.e., 2014, 2019, 2020), the maximum values observed during the growing season were used. AIRTREE simulations were conducted for each pixel of the study area, and for each year of study, we extracted the LAI values filtering out pixels classified as buildings, urban, wetlands, meadows and artificial settlements. LAI is a critical value for the proper evaluation of ecosystem services via AIRTREE model [14]. Therefore, we compared satellite data with field data observations collected with multiple simultaneous measurements of transmittance using the LAI-2200C Plant Canopy Analyzer (Li-Cor, Lincoln, NE, USA) and with a method based on the use of hemispherical photos collected in the frame of ICOS activities [60]. The sampling points were selected through a stratified sampling procedure (Figure S5) over the main four LCTs: Mediterranean maquis, Deciduous broadleaves, Pine forest and Holm oak forest (Table S2). For more details on the statistical design and the field campaigns for LAI measurement, please see the section "In situ LAI measurements" in the Supplementary File. To derive a function suitable to provide a realistic estimate of the annual dynamics of LAI for each LCT group, monthly values of LAI were fit to a set of parametric non-linear models (Gaussian, Exponential, Rational, Power, Sin, Weibull and Fourier). The function that best fitted data (lowest RMSE) was implemented in the model. The Fourier model was identified as the most appropriate model to reproduce LAI dynamics through the three years of study for the LCTs functional groups:

$$LAI_{norm} = a + b \cdot cos(x \cdot d) + c \cdot sin(x \cdot w) \quad (5)$$

where a, b, c and d were LCT-specific fixed values (i.e., the coefficients of the fitted model), and x is the day of the year (Figure 2). Coefficients of the Fourier models used for each LCT and for each of the three years of study are summarized in Table S3. Non-linear model intercomparisons are shown in Tables S4–S24.

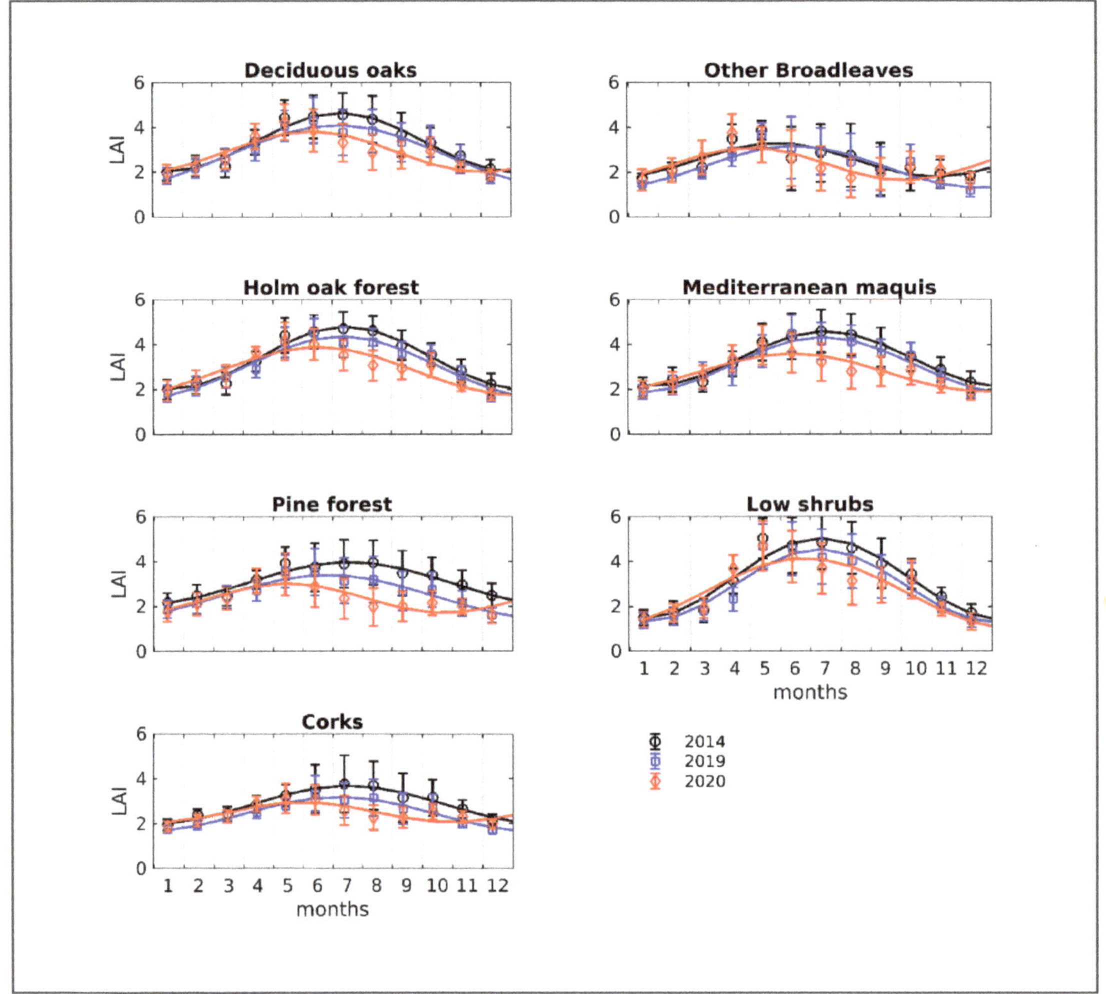

Figure 2. Fourier's model (lines) fitted to monthly LAI ($m^2\ m^{-2}$) satellite data (circles) for each LCT, for the three years of study. Each function was implemented into the AIRTREE model and used to estimate LAI dynamics during the year. Coefficients of the models are reported in Table S1.

Such a function was finally implemented into the AIRTREE model for each reference year in order to derive the daily LAI at each model time step. Canopy heights were only available for the year 2019 at a spatial resolution of 25 m [61]. These were resampled to 300 m in order to align the maps with those of LAI and canopy cover. All maps out on the boundaries of the Estate were cut, and pixels falling into the non-forested LCT were removed, so as to work only on the actual forested areas.

3. Results and Discussions

3.1. LAI and Canopy Cover Dynamics

Overall, average LAI for the entire Estate was 4.88, 4.31 and 3.90 in the 3 years of study, respectively (Figures S4 and S6), indicating a decrease of 11.83% and 20.05% in 2019 and 2020 compared to 2014, respectively. LAI showed a decreasing trend for all LCTs (Figure 3), in particular for Pine forest (for which the representability of the total LAI changes from 14% in 2014 to 12% in 2020, Figure S5). In 2020, LAI increased for low shrubs by only 1% when compared to 2019. This is also confirmed by canopy cover data (Figure S7) showing an average loss of 6.83% between 2014 and 2020. In particular, the Pine forest showed a loss of cover of 13.33%, of which 9.83% occurred in a year, between 2019 and 2020. Indeed, a positive linear correlation between changes in LAI and changes in canopy cover was observed (Pearson's $r = 0.63$). Although an increase in forest cover of 2% has been observed between 2010 and 2015 in the Mediterranean basin [62], the loss of Cover (7%) and LAI (up to 20%) experienced by the vegetation of the Estate between 2014 and 2020 confirms the increasing vulnerability of forests due to the climatic stressors and aridity the vegetation is exposed to [63]. Indeed, from 2014 (previously described wet year [30]), warmer and drier years succeeded. This may have caused stress and loss of productivity for Mediterranean forests, even if they are adapted to the typical dry and hot summers [64–67]. All ecosystems except shrubs showed a reduction in their LAI up to 14% (Figure 3).

The environmental monitoring data concerning the health status of the Castelporziano forests, detected from 2015 to today through the diachronic interpretation of the NDVI values provided by Sentinel-2 data, showed widespread suffering in pine and deciduous oaks forests [68]. Regarding the pine forests, two insect infestations occurred in 2015 and 2017, causing a widespread decline in the photosynthetic capacity (−34% in the average NDVI values for the period 2015–2021) of the forest, which caused the death of plants over large area [68]. In Deciduous oak forests, the limiting factor seems to be strongly correlated with summer drought and springs characterized by low amount of rain, causing a mean decrease in NDVI values of −27% for the observation period [37,68].

Interestingly, there is high resilience from several species of Evergreen shrubs, capable of adapting to drought stress even more than *Quercus ilex* has been previously observed [69,70], and therefore, a transformation of forest structure into a shrubland under increasing heatwaves and drought is the most likely scenario [64,71]. Indeed, Mediterranean plants have developed various morphological and physiological strategies to adapt to drought [72], but the largest trees are frequently more sensitive and less resistant and resilient to increases in aridity [72,73]. In line with our observations, Lloret et al. [74] showed that shorter trees are more resistant and resilient to increases in drought than taller trees. Tall shrubs such as *Phillyrea* sp., which are abundant in the Estate of Castelporziano, have been shown to have a higher capacity than trees to adapt and resist intensive droughts [75,76] and have a higher capacity than sympatric trees to maintain their foliage and concentrations of non-structural carbohydrates after droughts [76].

The inter-comparison between non-linear models fitted to monthly values of LAI for each year of study and for each LCT is shown in Tables S4–S24. On the 21 fits performed (i.e., 3 years for 7 LCTs), the Fourier function (Figure 3 and Table S3) was found to be the best model (highest R^2_{adj} and lowest RMSE) 12 times, although it produced some minor biases (i.e., increase in LAI for "other broadleaves" and "Corks" at the end of the year).

A general change in phenology was observed, in 2020 compared to 2014, with an anticipated peak in LAI that shifted from June to July in 2014 to May to June 2020 (Figure 3). This is in line with previous findings showing that key Mediterranean species such as *Quercus ilex* display a longer period of growth by approximately 10 days by advancing the onset of spring by winter warming, with an early cessation of growth in spring and summer [77].

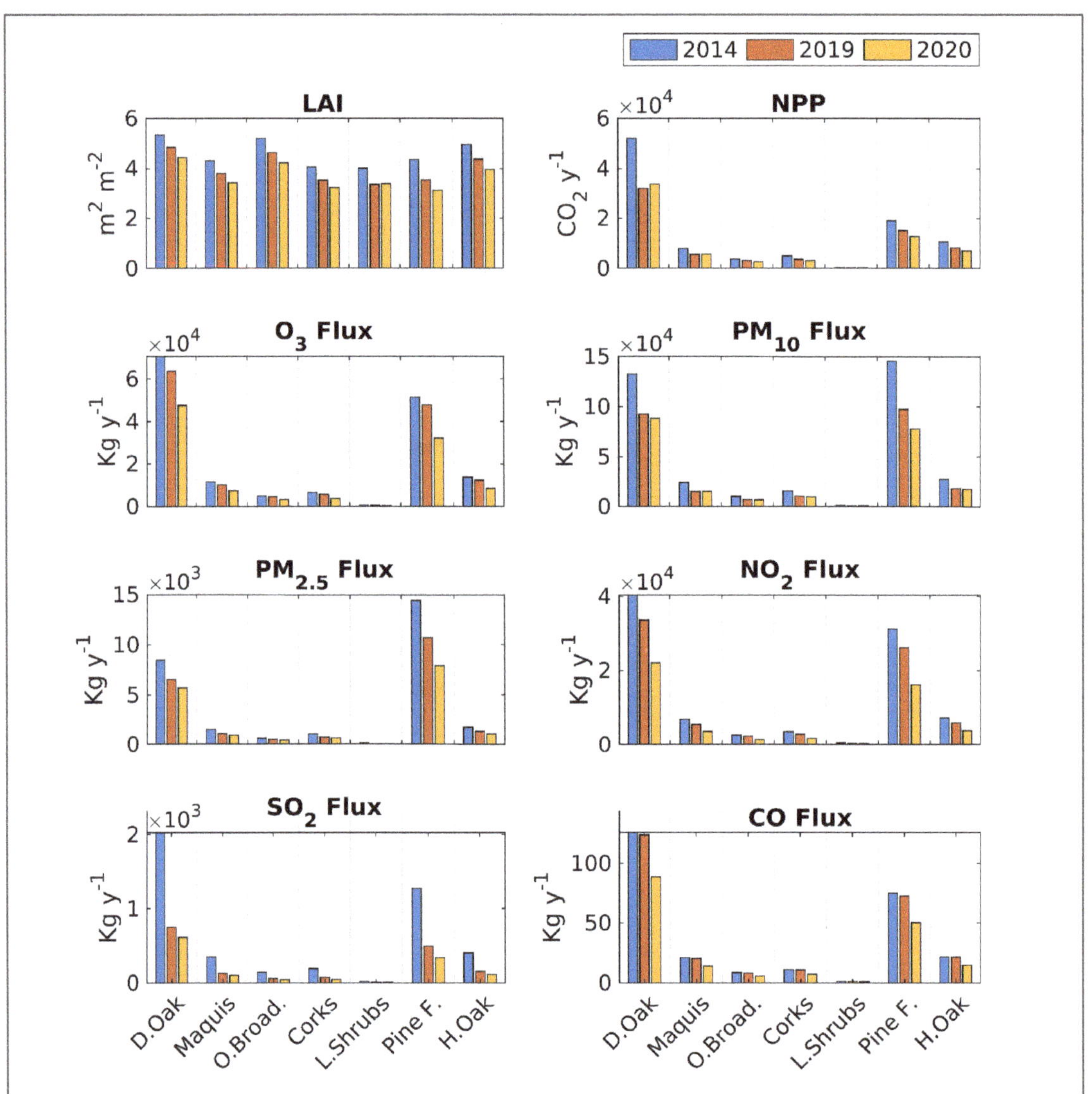

Figure 3. Intercomparison of the three years of study for each of the six Land Cover Types (Deciduous oaks, Mediterranean maquis, Other broadleaves, Corks, Low shrubs, Pine forest, Holm oak forest) used in this study. LAI (leaf area/ground area, m^2 m^{-2}), Net Primary Productivity (NPP, kg of CO_2 y^{-1}), deposition of ozone (kg O_3 y^{-1}), particulate matter (kg $PM_{2.5}$ y^{-1} and kg PM_{10} y^{-1}), Nitrogen dioxide (kg NO_2 y^{-1}), Sulphur dioxide (kg SO_2 y^{-1}) and Carbon monoxide (kg CO y^{-1}) estimated by the AIRTREE model are shown in the bar charts. Values represents cumulated values of uptake and deposition of pollutants for each LCT.

Satellite LAI values used in this study were compared with field measurements with an overestimation of 20% considering all LCTs (Figure 4). Deciduous broadleaves and Holm oak forests showed the highest LAI values (4.12 and 4.41, respectively; statistics are shown in Table S2), in accordance with what was previously found for Deciduous oaks stands [78]. We measured lower LAI values for Pine forests and Mediterranean maquis (2.08 and 3.85, respectively), in line with previous findings [79,80].

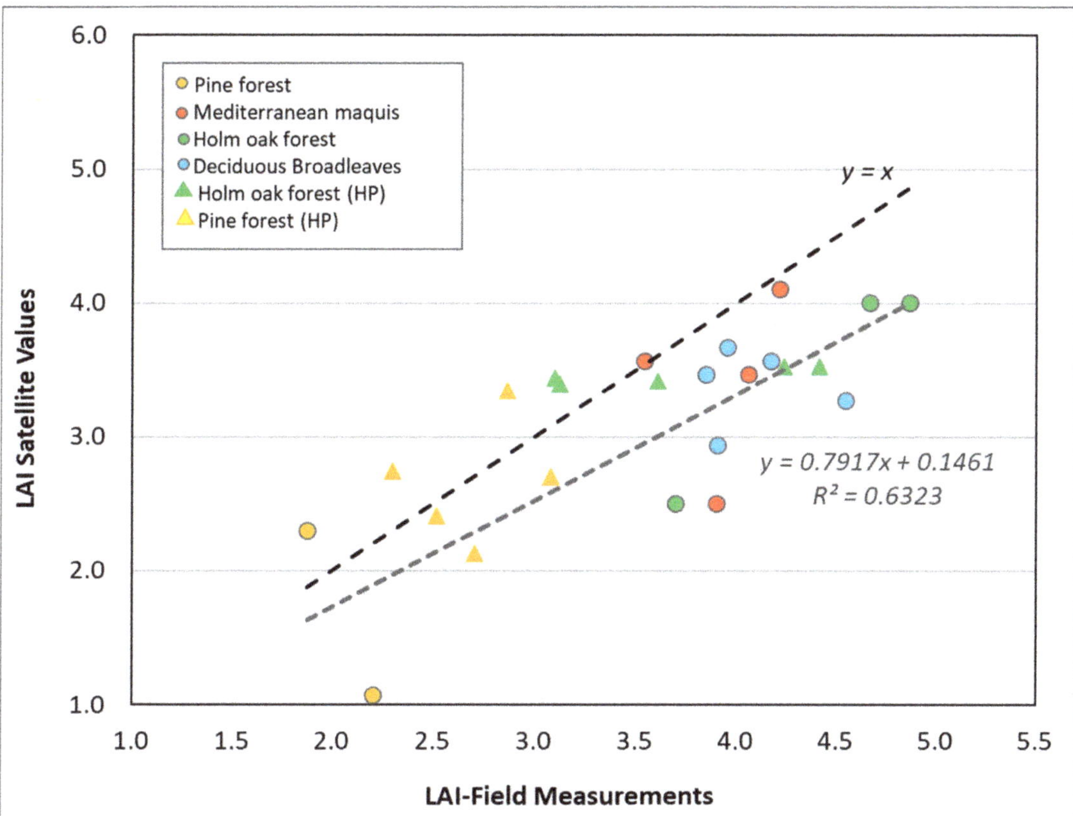

Figure 4. Correlation between ground-level measurement and satellite-level values of LAI ($m^2\ m^{-2}$); different colors represent different LCT sampled inside the Estate. The circles indicate field measurements carried out with LAI-2200C Plant Canopy Analyzer; triangles indicate field measurements carried out with hemispherical photography (HP).

3.2. Sequestration of Carbon Dioxide

Total yearly estimates of Net Primary Productivity (NPP) indicate that the Estate sequestrated 0.0981, 0.0675 and 0.0648 mt $CO_2\ y^{-1}$ for the years 2014, 2019 and 2020, respectively (Figures 2 and S8). In particular, Deciduous Oaks, Pine forests and Holm oak forests were the LCTs that contributed most: Deciduous Oaks contributed up to 53% (0.052 mt y^{-1}), Pine forest up to 22% (0.019 mt y^{-1}) and Holm oak forests up to 12% (0.01 mt y^{-1}) (Figure S9). The contribution to total carbon uptake by the other ecosystems (Other broadleaves 5%, corks 4%, Mediterranean maquis 9% and low shrubs <1%) was below 20%.

Compared to 2014, the Estate showed a decrease in productivity by 31.6% and 34.3% in 2019 and 2020, respectively (Figures 5a and S10). Between 2019 and 2020, Deciduous Oak and Mediterranean maquis and Low shrubs showed an increase in carbon uptake by 5.7%, 2.6% and 22.5%, respectively (in line with our previous statements about highest tolerance to drought stress).

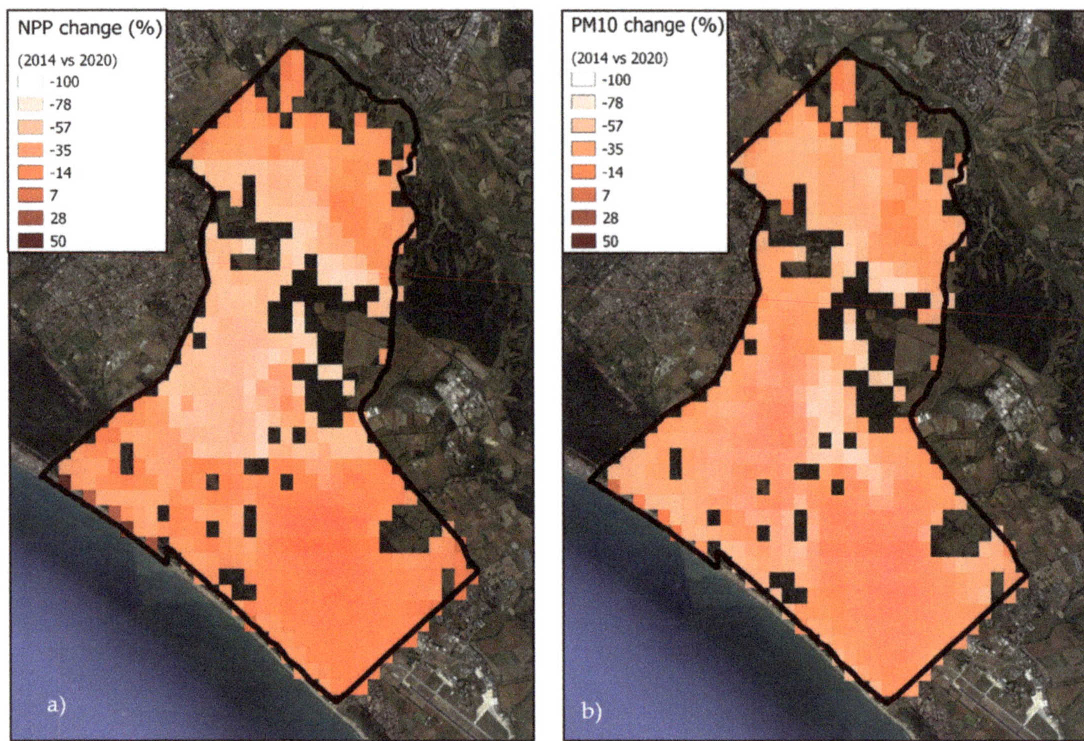

Figure 5. Percent changes (%) in (**a**) NPP (kg CO_2 y^{-1}) and (**b**) PM_{10} deposition (kg PM y^{-1}) estimated by the AIRTREE model between 2014 and 2020. The vividness of the pixels (from dark red to white) indicates the increased or decreased sequestration capacity of the Estate, while empty (transparent) pixels represent non-natural LCT not considered in this study.

Considering that pollutants such as ozone have been found to reduce carbon assimilation in the Holm oak forest up to 10% [14], we do not exclude a possible beneficial effect of reducing atmospheric pollutants in 2020 due to lockdown status in the first part of the year. Other LCTs instead, showed a decrease in carbon uptake up to 17%, and such decrease can be highly appreciated by comparing 2014 and 2020 (Figure S9).

We explain such behavior considering the different temperature recorded during the warm seasons in these two dry years. Otherwise, in 2019, the average temperature during summer was higher than in 2020 (Figure S2), and accordingly, the high vapor pressure deficit may have caused stomatal closure and reduced photosynthesis, as previously discussed in Conte et al. [30]. Looking at single LCT, the data show (Table 1) that broadleaf species and Holm oaks are the ecosystems with the higher NPP, with values up to 2.46 kg CO_2 m^2 y^{-1} in the wet year (2014), with abrupt decreases in productivity in the dry years. These model results, especially for Holm oak, are in line with data measured with Eddy Covariance at the ICOS Holm oak It-Cp2 site at the Estate (data not shown). Like Italy, warmer and drier conditions linked to increases in Atlantic multidecadal oscillations are associated with the increase in post-1990 defoliation in the forests of Spain [81]. Our results are in agreement with previous studies that have clearly indicated that the decreases such as dieback, defoliation and lower growth in Mediterranean oaks (*Quercus* spp.) and pines (*Pinus* spp.) in southern Europe are mainly due to more frequent drought, often interacting with higher temperatures (higher water demand) and pathogenic attack [82–86].

Such a cumulative effect of stressors has been associated with pathogen infestations in Mediterranean forests [86–89].

Table 1. Mean annual estimate and standard deviation of Carbon uptake (NPP), Ozone (O_3 flux), Particulate matter fluxes (PM_{10} and $PM_{2.5}$), fluxes of Nitrogen dioxide (NO_2), Sulphur dioxide (SO_2) and Carbon monoxide (CO) for the three years of study (2014, 2019 and 2020), for each LCT are reported.

LCT	Year	NPP	O_3 Flux	PM_{10} Flux	$PM_{2.5}$ Flux	NO_2 Flux	SO_2 Flux	CO Flux
		(Kg CO^2 m^2 y^{-1})	(g m^2 y^{-1})	(g m^2 y^{-1})	(g m^2 y^{-1})	(g m^2 y^{-1})	(g m^2 y^{-1})	(g m^2 y^{-1})
Low shrubs	2014	1.53 ± 0.44	2.52 ± 0.30	5.78 ± 0.27	0.36 ± 0.02	1.60 ± 0.24	0.08 ± 0.013	0.005 ± 0.0003
	2019	0.74 ± 0.26	2.22 ± 0.07	3.26 ± 0.40	0.23 ± 0.06	1.22 ± 0.02	0.03 ± 0.003	0.005 ± 0.0003
	2020	0.94 ± 0.09	1.65 ± 0.12	3.97 ± 0.64	0.25 ± 0.04	0.84 ± 0.05	0.02 ± 0.003	0.004 ± 0.0002
Other broadleaves	2014	2.46 ± 0.31	3.23 ± 0.19	6.69 ± 0.58	0.42 ± 0.04	1.79 ± 0.11	0.10 ± 0.006	0.006 ± 0.0002
	2019	2.10 ± 0.38	3.14 ± 0.13	4.72 ± 0.59	0.34 ± 0.04	1.60 ± 0.05	0.04 ± 0.003	0.006 ± 0.0002
	2020	1.81 ± 0.20	2.32 ± 0.12	4.63 ± 0.61	0.29 ± 0.04	1.05 ± 0.04	0.03 ± 0.002	0.004 ± 0.0001
Holm oak forest	2014	2.45 ± 0.43	3.17 ± 0.32	6.35 ± 0.96	0.40 ± 0.06	1.69 ± 0.18	0.09 ± 0.012	0.005 ± 0.0002
	2019	1.91 ± 0.54	2.91 ± 0.32	4.27 ± 0.80	0.31 ± 0.07	1.41 ± 0.11	0.04 ± 0.005	0.005 ± 0.0002
	2020	1.75 ± 0.38	2.11 ± 0.30	4.28 ± 0.83	0.27 ± 0.05	0.95 ± 0.11	0.03 ± 0.005	0.004 ± 0.0002
Mediterranean maquis	2014	1.86 ± 0.65	2.74 ± 0.34	5.61 ± 1.72	0.35 ± 0.11	1.66 ± 0.21	0.08 ± 0.015	0.005 ± 0.0003
	2019	1.32 ± 0.57	2.56 ± 0.26	3.57 ± 1.31	0.26 ± 0.09	1.38 ± 0.10	0.03 ± 0.005	0.005 ± 0.0003
	2020	1.43 ± 0.36	1.95 ± 0.17	3.78 ± 1.20	0.23 ± 0.07	0.96 ± 0.07	0.03 ± 0.004	0.004 ± 0.0002
Pine forest	2014	2.09 ± 0.38	5.62 ± 0.35	15.78 ± 2.72	1.56 ± 0.27	3.42 ± 0.20	0.14 ± 0.016	0.008 ± 0.0001
	2019	1.69 ± 0.49	5.43 ± 0.43	10.89 ± 2.45	1.20 ± 0.27	2.96 ± 0.15	0.06 ± 0.008	0.008 ± 0.0001
	2020	1.54 ± 0.39	4.02 ± 0.41	9.52 ± 2.35	0.96 ± 0.24	2.03 ± 0.16	0.04 ± 0.008	0.006 ± 0.0001
Deciduous oaks	2014	2.36 ± 0.61	3.20 ± 0.42	5.95 ± 0.90	0.38 ± 0.06	1.83 ± 0.22	0.09 ± 0.014	0.006 ± 0.0002
	2019	1.43 ± 0.68	2.92 ± 0.26	4.21 ± 0.84	0.30 ± 0.06	1.55 ± 0.08	0.03 ± 0.005	0.006 ± 0.0002
	2020	1.55 ± 0.36	2.21 ± 0.23	4.05 ± 0.79	0.26 ± 0.05	1.03 ± 0.07	0.03 ± 0.005	0.004 ± 0.0002
Corks	2014	2.03 ± 0.50	2.66 ± 0.30	6.60 ± 1.41	0.45 ± 0.10	1.47 ± 0.17	0.08 ± 0.012	0.005 ± 0.0004
	2019	1.52 ± 0.32	2.42 ± 0.25	4.33 ± 0.82	0.31 ± 0.07	1.21 ± 0.10	0.03 ± 0.004	0.005 ± 0.0004
	2020	1.41 ± 0.37	1.70 ± 0.28	4.48 ± 0.81	0.31 ± 0.06	0.81 ± 0.11	0.02 ± 0.005	0.003 ± 0.0003

3.3. Sequestration of Pollutants

Total yearly estimates of ozone deposition (O_3) indicate that the Estate sequestrated 159, 145 and 103 t O_3 y^{-1} for the years 2014, 2019 and 2020, respectively (Figures 2 and S8). Deciduous Oaks, Pine forests and Holm oak forests were the LCTs that contributed most. Deciduous Oaks contributed up to 46% to total ozone deposition (70 t y^{-1}) and Pine forest up to 33% (51 t y^{-1}). The contribution to total ozone deposition by the other ecosystems (Holm oak forest up to 9%, Mediterranean maquis up to 7%, Other broadleaves up to 3%, corks up to 4% and low shrubs <1%) is below 25% (Figures S11 and S12). The modelled ozone deposition on Holm oak agrees with our previous measured with the Eddy Covariance carried out in 2013 and 2014 [43].

Compared to 2014, the Estate showed a decrease in ozone deposition by 8.8% and 35.2% in 2019 and 2020, respectively. In particular, Pine forest, Deciduous oaks, Other broadleaves and Corks showed a decrease in ozone uptake by 32.7%, 25.2%, 29.4% and 35.3%, respectively. Such a decrease in deposition (Figure 3), in part due to the loss of canopy cover and in part due to physiological limitations as demonstrated by NPP decreases (Table 1), suggests that land use changes and environmental stressors heavily compromise the capacity of the Estate and, in general, Mediterranean peri-urban forests. We must point out, however, that most of the decreases in 2020 compared with 2014 are due to a strong reduction in atmospheric ozone concentration due to the lockdown status in the first part of the year. Nevertheless, ozone concentrations were similar between 2014 and 2019, highlighting evidence of stomatal limitation to ozone deposition in the dry year as stressed in our previous work [26,90].

Total yearly estimates of particulate matter (PM) indicate that the Estate sequestrated 357, 241 and 215 t PM_{10} y^{-1} and 27, 21, and16 t $PM_{2.5}$ y^{-1} for the years 2014, 2019 and 2020, respectively (Figures S8 and S13a,b). Deciduous oaks and Pine forests were the LCTs that contributed equally to PM_{10} deposition (up to 41% Pine forest in 2014 and Deciduous oaks in 2020, Figure S13), while a higher contribution (up to 52%) of Pine forest was found for $PM_{2.5}$ (Figure S13). The contribution to total PM deposition by the other ecosystems is below 25% (Other broadleaves up to 3%, Corks up to 4%, Mediterranean maquis up to 6%, Holm oak forest 8% and low shrubs <1%) (Figure S13).

Compared to 2014, the Estate showed a decrease in PM_{10} deposition by 32.5% in 2019 and 39.6% in 2020 (Figure 5b), while there was a decrease in $PM_{2.5}$ deposition of 25% in 2019 and 40% in 2020. Between 2019 and 2020, a decrease of 10.5% and 19.9% was observed for PM_{10} and $PM_{2.5}$, respectively. In particular, the Pine forest showed a reduction of 19.8% and 26% for PM_{10} and $PM_{2.5}$ in 2020, respectively (Figures 2, S14 and S15). The same conclusions drawn for ozone work here, with a significant reduction in PM emissions during the lockdown in 2020 (Figure S2). To note, several hectares (0.7 km^2 in 2017 and 2022) of forests have been cut in 2017 and 2020 (and another 1 km^2 is going to be cut in the near future) probably due to a strong infection of *Tomicus destruens* and *Toumayella parvicornis* on *Pinus pinea* stands, as visible in the white pixels in Figure S13. Moreover, as shown in Table 1 (and also found elsewhere by Fares et al. [10]), individual performances in PM removal are higher in Pine forests; therefore, a loss of these forest stands represents a major loss in RES provision. This loss is particularly relevant for cities such as Rome, where *Pinus pinea* is a key tree of the urban landscape.

Concerning Nitrogen Dioxide (NO_2), we found values of 92, 76 and 49 t NO_2 y^{-1} sequestration for the years 2014, 2019 and 2020, respectively (Figures 2 and S8). Deciduous Oaks removed up to 45%, and Pine forests removed up to 34%. Mediterranean maquis and holm oak forest contribution to total fluxes were constant (up to 8%) in the three years of study (Figure S16). The contribution to total NO_2 deposition by the other ecosystems (Other broadleaves up to 3%, Corks up to 4% and Low shrubs < 1%) was below 10% (Figure S16). Compared to 2014, the Estate showed a decrease in NO_2 deposition of 17% and 46.8% in 2019 and 2020, respectively. In particular, the highest decreases in NO_2 deposition were observed for Holm oak forest (up to 36.2%) and Corks (up to 38%), Pine forest (37.9%) and Other broadleaves (36.9%). Similar to ozone, the same conclusions can be drawn for NO_2, since 2020 recorded much lower emissions due to lockdown status (Figures S2 and S17).

As a minor entity, the Estate sequestrated other pollutants such as sulfur dioxide (SO_2) and carbon monoxide (CO). In agreement with the removal dynamics observed for the other pollutants, a strong decrease was observed, with values decreasing from 4.4 to 1.6 and 1.26 t SO_2 y^{-1} and from 0.26 to 0.25 and 0.18 t CO y^{-1} for the years 2014, 2019 and 2020, respectively (Figure 3).

4. Conclusions

The Mediterranean Basin is a global hotspot of biological diversity and the most diverse biome in Europe. However, biotic and abiotic stressors can compromise survival of native ecosystems, especially in highly urbanized contexts. This motivated us to investigate which environmental factors affect Mediterranean forests' health status and whether natural and anthropogenic stressors can have an impact on RES. Our study is the result of the integration of remote sensing products with a mechanistic modelling approach to estimate plant functionality and stress response. Despite previous results showing that an increase in forest cover by 2% has been observed between 2010 and 2015 in the Mediterranean region [91], we highlight that in response to climatic changes, pollution and biotic stresses, not only the extensiveness of peri-urban forests can be reduced (we described a loss of canopy cover by 7% and of LAI up to 20% between 2014 and 2020) but also their capacity to deliver RES. Our results also warn that future forest composition may be altered with an increase in Mediterranean shrubs in place of forests stands populated by pines and oaks.

The obtained results can have important implications for future forest management, and to raise awareness on the high ecological and economic value of RES. In this context, the quantitative estimates of RES may play a key role in supporting decision makers and management planners to evaluate possible future impacts of different practices or environmental policies.

Supplementary Materials: The following supporting information can be downloaded at: https://www.mdpi.com/article/10.3390/f13050689/s1, Figure S1: Soil classification for the Castelporziano natural reserve derived from the Harmonized World Soil Database viewer (HWSD v.1.2) is reported. Figure S2: Meteorological inter-comparison between the three years of study seasonal values of mean temperatures, and cumulated precipitation (top left and top right, respectively) are shown together with average concentrations of PM, NO2, SO2 and CO. Figure S3: Spatial association of meteorological stations to homogeneous portions of the natural reserve. Yellow dots indicate the position of each one of the meteorological stations. Climatic conditions of the portions of the Estate (red shapes) were associated to each meteorological station. Figure S4: Superimposed maps of vegetation (colored shapes) to LAI (black and white pixels). Colored shapes represent homogeneous groups of vegetation (LCTs). Figure S5: Castelporziano map, showing the sampling points chosen randomly using QGis. Figure S6: Average values of LAI for the entire Estate of Castelporziano are reported for each of the three years of study (top). Relative contribution to overall LAI by each LCT is shown in pie charts (bottom), for 2014, 2019, and 2020, respectively. Figure S7: Bar plot showing the percent change of canopy cover values from satellite data between 2014 and 2019, in blue, and 2020 in yellow. Figure S8: Total estimates of LAI, NPP, and pollutants deposition for the whole Estate in the three years of study. Figure S9: Cumulated values of NPP for the entire Estate of Castelporziano are reported for each of the three years of study (top). Figure S10: Superimposed maps of vegetation (colored shapes) to NPP (black and white pixels). Figure S11: Superimposed maps of vegetation (colored shapes) to O3 fluxes (black and white pixels). Figure S12: Cumulated values of O3 deposition for the entire Estate of Castelporziano are reported for each of the three years of study (top). Figure S13: Superimposed maps of vegetation (colored shapes) to PM10 (left) and PM2.5 (right) deposition on canopies (black and white pixels). Figure S14: Cumulated values of PM10 deposition for the entire Estate of Castelporziano are reported for each of the three years of study (top). Figure S15: Cumulated values of PM2.5 deposition for the entire Estate of Castelporziano are reported for each of the three years of study (top). Figure S16: Cumulated values of NO2 deposition for the entire Estate of Castelporziano are reported for each of the three years of study (top). Figure S17: Superimposed maps of vegetation (colored shapes) to NO2 deposition on canopies (black and white pixels). Table S1: For each LCT, the corresponding association is reported. For each association, average ecophysiological parameters (e.g. Vcmax and ozone tolerance) of the dominant species are used as model input to characterize the LCT ideal tree-type. Table S2: LAI, (m2m-2) measured for the different LCTs in Castelporziano Estate. Table S3: Coefficient of the Fourier model used in this study to simulate LAI at each model time-step. For each LCT, yearly coefficient that best (lowest RMSE) fit LAI measured by satellite data are shown. Table S4: Intercomparison between nonlinear models used to fit the day of the year (doy) and LAI of Corks for the year 2014. Table S5: Intercomparison between nonlinear models used to fit the day of the year (doy) and LAI of Other Broadleaves for the year 2014. Table S6: Intercomparison between nonlinear models used to fit the day of the year (doy) and LAI of Deciduous Oaks for the year 2014. Table S7: Intercomparison between nonlinear models used to fit the day of the year (doy) and LAI of Holm oak forest for the year 2014. Table S8: Intercomparison between nonlinear models used to fit the day of the year (doy) and LAI of Mediterranean maquis for the year 2014. Table S9: Intercomparison between nonlinear models used to fit the day of the year (doy) and LAI of Pine forest for the year 2014. Table S10: Intercomparison between nonlinear models used to fit the day of the year (doy) and LAI of Low shrubs for the year 2014. Table S11: Intercomparison between nonlinear models used to fit the day of the year (doy) and LAI of Other broadleaves for the year 2019. Table S12: Intercomparison between nonlinear models used to fit the day of the year (doy) and LAI of Corks for the year 2019. Table S13: Intercomparison between nonlinear models used to fit the day of the year (doy) and LAI of Deciduous oaks for the year 2019. Table S14: Intercomparison between nonlinear models used to fit the day of the year (doy) and LAI of Holm oak forest for the year 2019. Table S15: Intercomparison between nonlinear models used to fit the day of the year (doy) and LAI of Mediterranean maquis for the year 2019. Table S16: Intercomparison between nonlinear models used to fit the day of the year (doy) and LAI of Pine forest for the year 2019. Table S17: Intercomparison between nonlinear models used to fit the day of the year (doy) and LAI of Low shrubs for the year 2019. Table S18: Intercomparison between nonlinear models used to fit the day of the year (doy) and LAI of Other broadleaves for the year

2020. Table S19: Intercomparison between nonlinear models used to fit the day of the year (doy) and LAI of Corks for the year 2020. Table S20: Intercomparison between nonlinear models used to fit the day of the year (doy) and LAI of Deciduous oaks for the year 2020. Table S21: Intercomparison between nonlinear models used to fit the day of the year (doy) and LAI of Holm oak forest for the year 2020. Table S22: Intercomparison between nonlinear models used to fit the day of the year (doy) and LAI of Mediterranean maquis for the year 2020. Table S23: Intercomparison between nonlinear models used to fit the day of the year (doy) and LAI of Pine forest for the year 2020. Table S24: Intercomparison between nonlinear models used to fit the day of the year (doy) and LAI of Low shrubs for the year 2020.

Author Contributions: A.C. and I.Z. equally contributed to the work; conceptualization, S.F., A.C. and I.Z.; methodology, F.R., A.A., L.F., S.F., I.Z. and A.C.; validation, S.F., A.C., I.Z. and L.F.; investigation, I.Z., L.F., S.F., T.S. and V.M.; data curation T.S. and A.C.; writing—original draft preparation, all authors; supervision, S.F.; funding acquisition, S.F. All authors have read and agreed to the published version of the manuscript.

Funding: UE LIFE programme financial instrument (LIFE18 PRE IT 003) VEG-GAP "Vegetation for Urban Green Air Quality Plans"; the project funded by the Ministry of Research n. 20173RRN2S_001 PRIN 2017-EUFORICC "Establishing Urban FORest based solutions In Changing Cities"; PRIN 2020-MULTIFOR "Multi-scale observations to predict Forest response to pollution and climate change"; the project funded by Regione Lazio n. 36388 TECNOVERDE: "Tecnologie geomatiche e ambientali di precisione per il monitoraggio e la valorizzazione dei servizi ecosistemici delle infra- strutture verdi urbane e peri-urbane"; the 2021 @CNR project funded by the National Research Council BIOCITY "Riforestazione urbana: nuovi strumenti conoscitivi e di supporto decisionale".

Informed Consent Statement: Informed consent was obtained from all subjects involved in the study.

Acknowledgments: The research was made possible thanks to the Directorate and the Scientific commission of Castelporziano Estate.

Conflicts of Interest: The authors declare no conflict of interest.

References

1. Reid, W.V.; Mooney, H.A.; Cropper, A.; Capistrano, D.; Carpenter, S.R.; Chopra, K.; Dasgupta, P.; Dietz, T.; Duraiappah, A.K.; Hassan, R.; et al. Millennium ecosystem assessment. In *Ecosystems and Human Well-Being*; Island Press United States of America: Washington, DC, USA, 2005; Volume 5.
2. Bäckstrand, K.; Lövbrand, E. Planting Trees to Mitigate Climate Change: Contested Discourses of Ecological Modernization, Green Governmentality and Civic Environmentalism. *Glob. Environ. Politics* **2006**, *6*, 50–75. [CrossRef]
3. Mengist, W.; Soromessa, T.; Feyisa, G.L. A global view of regulatory ecosystem services: Existed knowledge, trends, and research gaps. *Ecol. Process.* **2020**, *9*, 40. [CrossRef]
4. Villamagna, M.; Angermeier, P.L.; Bennett, E.M. Capacity, pressure, demand, and flow: A conceptual framework for analyzing ecosystem service provision and delivery. *Ecol. Complex.* **2013**, *15*, 114–121. [CrossRef]
5. Bernstein, L.; Bosch, P.; Canziani, O.; Chen, Z.; Christ, R.; Riahi, K. *IPCC, 2007: Climate Change 2007: Synthesis Report*; IPCC: Geneva, Switzerland, 2008.
6. Otu-Larbi, F.; Conte, A.; Fares, S.; Wild, O.; Ashworth, K. Current and future impacts of drought and ozone stress on Northern Hemisphere forests. *Glob. Chang. Biol.* **2020**, *26*, 6218–6234. [CrossRef]
7. Wang, Y.; Frankenberg, C. On the impact of canopy model complexity on simulated carbon, water, and solar-induced chlorophyll fluorescence fluxes. *Biogeosciences* **2022**, *19*, 29–45. [CrossRef]
8. ARPA Lazio, Air Quality Documental Section. Traffic Stations Description. 2016. Available online: http://www.arpalazio.net/main/aria/doc/RQA/inqRQA.php (accessed on 24 March 2016).
9. Deserti, M.; Savoia, E.; Cacciamani, C.; Golinelli, M.; Kerschbaumer, A.; Leoncini, G.; Selvini, A.; Paccagnella, T.; Tibaldi, S. Operational meteorological pre-processing at Emilia-Romagna ARPA meteorological service as a part of a decision support system for air quality management. *Int. J. Environ. Pollut.* **2001**, *16*, 571–582. [CrossRef]
10. Fares, S.; Conte, A.; Alivernini, A.; Chianucci, F.; Grotti, M.; Zappitelli, I.; Petrella, F.; Corona, P. Testing Removal of Carbon Dioxide, Ozone, and Atmospheric Particles by Urban Parks in Italy. *Environ. Sci. Technol.* **2020**, *54*, 14910–14922. [CrossRef]
11. Haylock, M.R.; Hofstra, N.; Tank, A.M.G.K.; Klok, E.J.; Jones, P.D.; New, M. A European daily high-resolution gridded data set of surface temperature and precipitation for 1950–2006. *J. Geophys. Res. Atmos.* **2008**, *113*, 20119. [CrossRef]
12. Holloway, T.; Miller, D.; Anenberg, S.; Diao, M.; Duncan, B.; Fiore, A.M.; Henze, D.K.; Hess, J.; Kinney, P.L.; Liu, Y.; et al. Satellite Monitoring for Air Quality and Health. *Annu. Rev. Biomed. Data Sci.* **2021**, *4*, 417–447. [CrossRef]

13. Bonan, G.B.; Patton, E.G.; Finnigan, J.J.; Baldocchi, D.D.; Harman, I.N. Moving beyond the incorrect but useful paradigm: Reevaluating big-leaf and multilayer plant canopies to model biosphere-atmosphere fluxes—A review. *Agric. For. Meteorol.* **2021**, *306*, 108435. [CrossRef]
14. Fares, S.; Alivernini, A.; Conte, A.; Maggi, F. Ozone and particle fluxes in a Mediterranean forest predicted by the AIRTREE model. *Sci. Total Environ.* **2019**, *682*, 494–504. [CrossRef] [PubMed]
15. World Health Organization. Air Quality Guidelines for Europe. 2000. Available online: https://apps.who.int/iris/handle/10665/107335 (accessed on 1 March 2022).
16. Delaria, E.R.; Place, B.K.; Liu, A.X.; Cohen, R.C. Laboratory measurements of stomatal NO_2 deposition to native California trees and the role of forests in the NO_x cycle. *Atmos. Chem. Phys.* **2020**, *20*, 14023–14041. [CrossRef]
17. Paoletti, E. Impact of ozone on Mediterranean forests: A review. *Environ. Pollut.* **2006**, *144*, 463–474. [CrossRef] [PubMed]
18. Fusaro, L.; Mereu, S.; Salvatori, E.; Agliari, E.; Fares, S.; Manes, F. Modeling ozone uptake by urban and peri-urban forest: A case study in the Metropolitan City of Rome. *Environ. Sci. Pollut. Res.* **2018**, *25*, 8190–8205. [CrossRef] [PubMed]
19. Hoshika, Y.; Paoletti, E.; Centritto, M.; Gomes, M.T.G.; Puértolas, J.; Haworth, M. Species-specific variation of photosynthesis and mesophyll conductance to ozone and drought in three Mediterranean oaks. *Physiol. Plant.* **2022**, *174*, e13639. [CrossRef] [PubMed]
20. Ball, J.T.; Woodrow, I.E.; Berry, J.A. A Model Predicting Stomatal Conductance and its Contribution to the Control of Photosynthesis under Different Environmental Conditions. In *Progress in Photosynthesis Research*; Springer: Dordrecht, The Netherlands, 1987; pp. 221–224. [CrossRef]
21. Medlyn, B.E.; Duursma, R.A.; Eamus, D.; Ellsworth, D.S.; Prentice, I.C.; Barton, C.V.M.; Crous, K.Y.; DE Angelis, P.; Freeman, M.; Wingate, L. Reconciling the optimal and empirical approaches to modelling stomatal conductance. *Glob. Chang. Biol.* **2011**, *17*, 2134–2144. [CrossRef]
22. Katul, G.G.; Palmroth, S.; Oren, R. Leaf stomatal responses to vapour pressure deficit under current and CO2-enriched atmosphere explained by the economics of gas exchange. *Plant Cell Environ.* **2009**, *32*, 968–979. [CrossRef]
23. Fares, S.; Conte, A.; Chabbi, A. Ozone flux in plant ecosystems: New opportunities for long-term monitoring networks to deliver ozone-risk assessments. *Environ. Sci. Pollut. Res.* **2018**, *25*, 8240–8248. [CrossRef]
24. Fuster, B.; Sánchez-Zapero, J.; Camacho, F.; García-Santos, V.; Verger, A.; Lacaze, R.; Weiss, M.; Baret, F.; Smets, B. Quality Assessment of PROBA-V LAI, fAPAR and fCOVER Collection 300 m Products of Copernicus Global Land Service. *Remote Sens.* **2020**, *12*, 1017. [CrossRef]
25. Kamenova, I.; Dimitrov, P. Evaluation of Sentinel-2 vegetation indices for prediction of LAI, fAPAR and fCover of winter wheat in Bulgaria. *Eur. J. Remote Sens.* **2020**, *54*, 89–108. [CrossRef]
26. Conte, A.; Otu-Larbi, F.; Alivernini, A.; Hoshika, Y.; Paoletti, E.; Ashworth, K.; Fares, S. Exploring new strategies for ozone-risk assessment: A dynamic-threshold case study. *Environ. Pollut.* **2021**, *287*, 117620. [CrossRef] [PubMed]
27. Otu-Larbi, F.; Conte, A.; Fares, S.; Wild, O.; Ashworth, K. FORCAsT-gs: Importance of Stomatal Conductance Parameterization to Estimated Ozone Deposition Velocity. *J. Adv. Modeling Earth Syst.* **2021**, *13*, e2021MS002581. [CrossRef]
28. Ratna, S.B.; Ratnam, J.V.; Behera, S.K.; Cherchi, A.; Wang, W.; Yamagata, T. The unusual wet summer (July) of 2014 in Southern Europe. *Atmos. Res.* **2017**, *189*, 61–68. [CrossRef]
29. Nimbus Web Climatologia Locale. Available online: http://www.nimbus.it/clima/2015/150114clima2014.htm (accessed on 24 March 2022).
30. Conte, A.; Fares, S.; Salvati, L.; Savi, F.; Matteucci, G.; Mazzenga, F.; Spano, D.; Sirca, C.; Marras, S.; Galvagno, M.; et al. Ecophysiological Responses to Rainfall Variability in Grassland and Forests Along a Latitudinal Gradient in Italy. *Front. For. Glob. Chang.* **2019**, *2*, 16. [CrossRef]
31. Donzelli, G.; Cioni, L.; Cancellieri, M.; Morales, A.L.; Suárez-Varela, M.M.M. The Effect of the Covid-19 Lockdown on Air Quality in Three Italian Medium-Sized Cities. *Atmosphere* **2020**, *11*, 1118. [CrossRef]
32. Collivignarelli, M.C.; Abbà, A.; Bertanza, G.; Pedrazzani, R.; Ricciardi, P.; Miino, M.C. Lockdown for CoViD-2019 in Milan: What are the effects on air quality? *Sci. Total Environ.* **2020**, *732*, 139280. [CrossRef]
33. Cameletti, M. The Effect of Corona Virus Lockdown on Air Pollution: Evidence from the City of Brescia in Lombardia Region (Italy). *Atmos. Environ.* **2020**, *239*, 117794. [CrossRef]
34. Gualtieri, G.; Brilli, L.; Carotenuto, F.; Vagnoli, C.; Zaldei, A.; Gioli, B. Quantifying road traffic impact on air quality in urban areas: A Covid19-induced lockdown analysis in Italy. *Environ. Pollut.* **2020**, *267*, 115682. [CrossRef]
35. Balasubramaniam, D.; Kanmanipappa, C.; Shankarlal, B.; Saravanan, M. Assessing the impact of lockdown in US, Italy and France–What are the changes in air quality? *Energy Sources Part A Recovery Util. Environ. Eff.* **2020**. [CrossRef]
36. Cucca, B.; Recanatesi, F.; Ripa, M.N. Evaluating the Potential of Vegetation Indices in Detecting Drought Impact Using Remote Sensing Data in a Mediterranean Pinewood. *Lect. Notes Comput. Sci.* **2020**, *12253*, 50–62. [CrossRef]
37. Recanatesi, F.; Giuliani, C.; Rossi, C.M.; Ripa, M.N. A Remote Sensing-Assisted Risk Rating Study to Monitor Pinewood Forest Decline: The Study Case of the Castelporziano State Nature Reserve (Rome). *Smart Innov. Syst. Technol.* **2019**, *100*, 68–75. [CrossRef]
38. della Rocca, B.; Pignatti, S.; Mugnoli, S.; Bianco, P.M. La carta della vegetazione della Tenuta di Castelporziano. *Accad. Naz. Delle Sci. Detta Dei XL Scr. E Doc.* **2001**, *26*, 709–748.

39. Davison, B.; Taipale, R.; Langford, B.; Misztal, P.; Fares, S.; Matteucci, G.; Loreto, F.; Cape, J.N.; Rinne, J.; Hewitt, C.N. Concentrations and fluxes of biogenic volatile organic compounds above a Mediterranean macchia ecosystem in western Italy. *Biogeosciences* **2009**, *6*, 1655–1670. [CrossRef]
40. Giordano, E.; Recanatesi, F.; Scrinzi, G. La Biodiversit Forestale Della Tenuta Presidenziale di Castelporziano. 2017. Available online: https://dspace.unitus.it/handle/2067/35638 (accessed on 1 March 2022).
41. Manes, F.; Grignetti, A.; Tinelli, A.; Lenz, R.; Ciccioli, P. General features of the Castelporziano test site. *Atmos. Environ.* **1997**, *31*, 19–25. Available online: https://www.researchgate.net/profile/Aleandro-Tinelli/publication/354031145_General_features_of_the_castelporziano_test_site_1997/links/611fba3d169a1a01031631f2/General-features-of-the-castelporziano-test-site-1997.pdf (accessed on 17 March 2022). [CrossRef]
42. Fares, S.; Mereu, S.; Mugnozza, G.S.; Vitale, M.; Manes, F.; Frattoni, M.; Ciccioli, P.; Gerosa, G.; Loreto, F. The ACCENT-VOCBAS field campaign on biosphere-atmosphere interactions in a Mediterranean ecosystem of Castelporziano (Rome): Site characteristics, climatic and meteorological conditions, and eco-physiology of vegetation. *Biogeosci. Discuss.* **2009**, *6*, 1185–1227. [CrossRef]
43. Fusaro, L.; Salvatori, E.; Mereu, S.; Silli, V.; Bernardini, A.; Tinelli, A.; Manes, F. Researches in Castelporziano test site: Ecophysiological studies on Mediterranean vegetation in a changing environment. *Rend. Lincei* **2015**, *26*, 473–481. [CrossRef]
44. Fares, S.; Savi, F.; Muller, J.; Matteucci, G.; Paoletti, E. Simultaneous measurements of above and below canopy ozone fluxes help partitioning ozone deposition between its various sinks in a Mediterranean Oak Forest. *Agric. For. Meteorol.* **2014**, *198*, 181–191. [CrossRef]
45. Fischer, G.; Nachtergaele, F.; Prieler, S.; van Velthuizen, H.T.; Verelst, L.; Wiberg, D. *Global Agro-Ecological Zones Assessment for Agriculture (GAEZ 2008)*; IIASA: Laxenburg, Austria; FAO: Rome, Italy, 2008; Volume 10.
46. Baldocchi, D. An analytical solution for coupled leaf photosynthesis and stomatal conductance models. *Tree Physiol.* **1994**, *14*, 1069–1079. [CrossRef]
47. Keenan, T.; Sabate, S.; Gracia, C. Soil water stress and coupled photosynthesis–conductance models: Bridging the gap between conflicting reports on the relative roles of stomatal, mesophyll conductance and biochemical limitations to photosynthesis. *Agric. For. Meteorol.* **2010**, *150*, 443–453. [CrossRef]
48. Lombardozzi, D.; Sparks, J.P.; Bonan, G. Integrating O3 influences on terrestrial processes: Photosynthetic and stomatal response data available for regional and global modeling. *Biogeosci. Discuss.* **2013**, *10*, 6973–7012. [CrossRef]
49. Lombardozzi, D.; Levis, S.; Bonan, G.; Sparks, J.P. Predicting photosynthesis and transpiration responses to ozone: Decoupling modeled photosynthesis and stomatal conductance. *Biogeosciences* **2012**, *9*, 3113–3130. [CrossRef]
50. Lombardozzi, D.L.; Levis, S.; Bonan, G.; Hess, P.G.; Sparks, J.P. The Influence of Chronic Ozone Exposure on Global Carbon and Water Cycles. *J. Clim.* **2015**, *28*, 292–305. [CrossRef]
51. Zhang, L.; Brook, J.R.; Vet, R. On ozone dry deposition—with emphasis on non-stomatal uptake and wet canopies. *Atmos. Environ.* **2002**, *36*, 4787–4799. [CrossRef]
52. Baldocchi, D.D.; Hicks, B.B.; Camara, P. A canopy stomatal resistance model for gaseous deposition to vegetated surfaces. *Atmos. Environ.* **1987**, *21*, 91–101. [CrossRef]
53. Pederson, J.; Massman, W.; Mahrt, L.; Delany, A.; Oncley, S.; Hartog, G.; Neumann, H.; Mickle, R.; Shaw, R.; Paw, K.T.; et al. California ozone deposition experiment: Methods, results, and opportunities. *Atmos. Environ.* **1995**, *29*, 3115–3132. [CrossRef]
54. Bidwell, R.G.S.; Fraser, D.E. Carbon monoxide uptake and metabolism by leaves. *Can. J. Bot.* **2011**, *50*, 1435–1439. [CrossRef]
55. Wesely, M.L. Parameterization of surface resistances to gaseous dry deposition in regional-scale numerical models. *Atmos. Environ.* **2007**, *41*, 52–63. [CrossRef]
56. Lovett, G.M. Atmospheric Deposition of Nutrients and Pollutants in North America: An Ecological Perspective. *Ecol. Appl.* **1994**, *4*, 629–650. [CrossRef]
57. Aromolo, R.; Moretti, V.; Sorgi, T. Setting Up Optimal Meteorological Networks: An Example From Italy. *Strateg. Plan. Energy Environ.* **2021**, *40*, 39–54. [CrossRef]
58. Recanatesi, F. Variations in land-use/land-cover changes (LULCCs) in a peri-urban Mediterranean nature reserve: The estate of Castelporziano (Central Italy). *Rend. Lincei* **2015**, *3*, 517–526. [CrossRef]
59. European Commission Directorate-General Joint Research Centre. Leaf Area Index: Version 1333 m Resolution, Globe, 10-Daily. 2017. Available online: http://land.copernicus.vgt.vito.be/geonetwork/srv/api/records/urn:cgls:global:lai300_v1_333m (accessed on 27 March 2022).
60. Gielen, B.; de Beeck, M.; Michilsens, M.; Papale, D. *ICOS Ecosystem Instructions for Ancillary Vegetation Measurements in Forest (Version 20200330)*; ICOS Ecosystem Thematic Centre: Viterbo, Italy, 2017. [CrossRef]
61. Potapov, P.; Li, X.; Hernandez-Serna, A.; Tyukavina, A.; Hansen, M.C.; Kommareddy, A.; Pickens, A.; Turubanova, S.; Tang, H.; Silva, C.E.; et al. Mapping global forest canopy height through integration of GEDI and Landsat data. *Remote Sens. Environ.* **2020**, *253*, 112165. [CrossRef]
62. At Risk Mediterranean Forests Make Vital Contributions to Development UN News. Available online: https://news.un.org/en/story/2018/11/1026761 (accessed on 24 March 2022).
63. Penuelas, J.; Gracia, C.; Jump, I.F.A.; Carnicer, J.; Coll, M.; Lloret, F.; Yuste, J.C.; Estiarte, M.; Rutishauser, T.; Ogaya, R. Introducing the climate change effects on Mediterranean forest ecosystems: Observation, experimentation, simulation and management. In *Forêt Méditerranéenne t. XXXI, n 4*, 4th ed.; Association Forêt Méditerranéenne: Marseille, France, 2010; pp. 357–362. Available online: http://hdl.handle.net/2042/39215 (accessed on 23 March 2022).

64. Sardans, J.; Peñuelas, J. Plant-soil interactions in Mediterranean forest and shrublands: Impacts of climatic change. *Plant Soil* **2013**, *365*, 1–33. [CrossRef] [PubMed]
65. Misson, L.; Degueldre, D.; Collin, C.; Rodriguez, R.; Rocheteau, A.; Ourcival, J.-M.; Rambal, S. Phenological responses to extreme droughts in a Mediterranean forest. *Glob. Chang. Biol.* **2011**, *17*, 1036–1048. [CrossRef]
66. Bongers, F.J.; Olmo, M.; Lopez-Iglesias, B.; Anten, N.P.R.; Villar, R. Drought responses, phenotypic plasticity and survival of Mediterranean species in two different microclimatic sites. *Plant Biol.* **2017**, *19*, 386–395. [CrossRef]
67. Castagneri, D.; Regev, L.; Boaretto, E.; Carrer, M. Xylem anatomical traits reveal different strategies of two Mediterranean oaks to cope with drought and warming. *Environ. Exp. Bot.* **2017**, *133*, 128–138. [CrossRef]
68. Recanatesi, F.; Giuliani, C.; Ripa, M.N. Monitoring Mediterranean Oak Decline in a Peri-Urban Protected Area Using the NDVI and Sentinel-2 Images: The Case Study of Castelporziano State Natural Reserve. *Sustainability* **2018**, *10*, 3308. [CrossRef]
69. Ogaya, R.; Peñuelas, J. Tree growth, mortality, and above-ground biomass accumulation in a holm oak forest under a five-year experimental field drought. *Plant Ecol.* **2006**, *189*, 291–299. [CrossRef]
70. Barbeta, A.; Peñuelas, J. Sequence of plant responses to droughts of different timescales: Lessons from holm oak (Quercus ilex) forests. *Plant Ecol. Divers.* **2016**, *9*, 321–338. [CrossRef]
71. Barbeta, A.; Mejía-Chang, M.; Ogaya, R.; Voltas, J.; Dawson, T.E.; Peñuelas, J. The combined effects of a long-term experimental drought and an extreme drought on the use of plant-water sources in a Mediterranean forest. *Glob. Chang. Biol.* **2015**, *21*, 1213–1225. [CrossRef]
72. Medrano, H.; Flexas, J.; Galmés, J. Variability in water use efficiency at the leaf level among Mediterranean plants with different growth forms. *Plant Soil* **2009**, *317*, 17–29. [CrossRef]
73. Sperlich, D.; Chang, C.T.; Peñuelas, J.; Gracia, C.; Sabaté, S. Seasonal variability of foliar photosynthetic and morphological traits and drought impacts in a Mediterranean mixed forest. *Tree Physiol.* **2015**, *35*, 501–520. [CrossRef] [PubMed]
74. Lloret, F.; Siscart, D.; Dalmases, C. Canopy recovery after drought dieback in holm-oak Mediterranean forests of Catalonia (NE Spain). *Glob. Chang. Biol.* **2004**, *10*, 2092–2099. [CrossRef]
75. Rosas, T.; Galiano, L.; Ogaya, R.; Peñuelas, J.; Martínez-Vilalta, J. Dynamics of non-structural carbohydrates in three mediterranean woody species following long-term experimental drought. *Front. Plant Sci.* **2013**, *4*, 400. [CrossRef] [PubMed]
76. Sarris, D.; Koutsias, N. Ecological adaptations of plants to drought influencing the recent fire regime in the Mediterranean. *Agric. For. Meteorol.* **2014**, *184*, 158–169. [CrossRef]
77. Lempereur, M.; Limousin, J.-M.; Guibal, F.; Ourcival, J.-M.; Rambal, S.; Ruffault, J.; Mouillot, F. Recent climate hiatus revealed dual control by temperature and drought on the stem growth of Mediterranean Quercus ilex. *Glob. Chang. Biol.* **2017**, *23*, 42–55. [CrossRef]
78. Cutini, A.; Matteucci, G.; Mugnozza, G.S. Estimation of leaf area index with the Li-Cor LAI 2000 in deciduous forests. *For. Ecol. Manag.* **1998**, *105*, 55–65. [CrossRef]
79. Manes, F.; Anselmi, S.; Giannini, M.; Melini, S. Relationships between leaf area index (LAI) and vegetation indices to analyze and monitor Mediterranean ecosystems. *Int. Soc. Opt. Photonics* **2001**, *4171*, 328–335. [CrossRef]
80. Gratani, L.; Crescente, M.F. Map-Making of Plant Biomass and Leaf Area Index for Management of Protected Areas. *Aliso A J. Syst. Florist. Bot.* **2000**, *19*, 1–12. [CrossRef]
81. Sánchez-Salguero, R.; Camarero, J.J.; Grau, J.M.; De La Cruz, A.C.; Gil, P.M.; Minaya, M.; Cancio, F. Analysing Atmospheric Processes and Climatic Drivers of Tree Defoliation to Determine Forest Vulnerability to Climate Warming. *Forests* **2016**, *8*, 13. [CrossRef]
82. Peña-Gallardo, M.; Vicente-Serrano, S.M.; Camarero, J.J.; Gazol, A.; Sánchez-Salguero, R.; Domínguez-Castro, F.; El Kenawy, A.; Beguería-Portugés, S.; Gutiérrez, E.; de Luis, M.; et al. Drought Sensitiveness on Forest Growth in Peninsular Spain and the Balearic Islands. *Forests* **2018**, *9*, 524. [CrossRef]
83. Gea-Izquierdo, G.; Viguera, B.; Cabrera, M.; Cañellas, I. Drought induced decline could portend widespread pine mortality at the xeric ecotone in managed mediterranean pine-oak woodlands. *For. Ecol. Manag.* **2014**, *320*, 70–82. [CrossRef]
84. Braisier, M. Phytophthora cinnamomi and oak decline in southern Europe. Environmental constraints including climate change. *Ann. Des Sci. For.* **1996**, *53*, 347–358. [CrossRef]
85. Corcobado, T.; Cubera, E.; Juárez, E.; Moreno, G.; Solla, A. Drought events determine performance of Quercus ilex seedlings and increase their susceptibility to Phytophthora cinnamomi. *Agric. For. Meteorol.* **2014**, *192–193*, 1–8. [CrossRef]
86. Camarero, J.J.; Álvarez-Taboada, F.; Hevia, A.; Castedo-Dorado, F. Radial growth and wood density reflect the impacts and susceptibility to defoliation by gypsy moth and climate in radiata pine. *Front. Plant Sci.* **2018**, *871*, 1582. [CrossRef]
87. Mecheri, H.; Kouidri, M.; Boukheroufa-Sakraoui, F.; Adamou, A.E. Variation du taux d'infestation par Thaumetopoea pityocampa du pin d'Alep: Effet sur les paramètres dendrométriques dans les forêts de la région de Djelfa (Atlas saharien, Algérie). *Comptes Rendus Biol.* **2018**, *341*, 380–386. [CrossRef]
88. Avila, J.M.; Gallardo, A.; Ibáñez, B.; Gómez-Aparicio, L. Quercus suber dieback alters soil respiration and nutrient availability in Mediterranean forests. *J. Ecol.* **2016**, *104*, 1441–1452. [CrossRef]
89. Erkan, N. Impact of pine processionary moth (Thaumetopoea wilkinsoni Tams) on growth of Turkish red pine (Pinus brutia Ten.). *Afr. J. Agric. Res.* **2011**, *6*, 4983–4988. [CrossRef]

90. Savi, F.; Savi, F.; Fares, S. Ozone dynamics in a Mediterranean Holm oak forest: Comparison among transition periods characterized by different amounts of precipitation. *Ann. Silvic. Res.* **2014**, *38*, 1–6. [CrossRef]
91. Peñuelas, J.; Sardans, J. Global Change and Forest Disturbances in the Mediterranean Basin: Breakthroughs, Knowledge Gaps, and Recommendations. *Forests* **2021**, *12*, 603. [CrossRef]

Article

Defoliation, Recovery and Increasing Mortality in Italian Forests: Levels, Patterns and Possible Consequences for Forest Multifunctionality

Filippo Bussotti [1], Giancarlo Papitto [2], Domenico Di Martino [2], Cristiana Cocciufa [2], Claudia Cindolo [2], Enrico Cenni [1], Davide Bettini [1], Giovanni Iacopetti [1] and Martina Pollastrini [1,*]

[1] Department of Agriculture, Food, Environment and Forestry (DAGRI), University of Firenze, 50142 Firenze, Italy; filippo.bussotti@unifi.it (F.B.); enrico.cenni@unifi.it (E.C.); davide.bettini@unifi.it (D.B.); giovanni.iacopetti@gmail.com (G.I.)

[2] Arma dei Carabinieri, Comando per la Tutela della Biodiversità e dei Parchi, 00187 Roma, Italy; giancarlo.papitto@carabinieri.it (G.P.); dodimartino@gmail.com (D.D.M.); cristianacocciufa@gmail.com (C.C.); clacindol@gmail.com (C.C.)

* Correspondence: martina.pollastrini@unifi.it; Tel.: +39-055-275-5582

Abstract: Forest health and multifunctionality are threatened by global challenges such as climate change. Forest health is currently assessed within the pan-European ICP Forests (International Co-operative Programme on Assessment and Monitoring of Air Pollution Effects on Forests) programme through the evaluation of tree crown conditions (defoliation). This paper analyses the results of a 24-year assessment carried out in Italy on 253 permanent plots distributed across the whole forested area. The results evidenced a substantial stability of crown conditions at the national level, according to the usual defoliation thresholds Defoliation > 25% and Defoliation > 60%, albeit with species-specific patterns. Within this apparent temporal stability, an increased fraction of extremely defoliated and dead trees was observed. Extreme defoliation mostly occurred in years with severe summer drought, whereas mortality was higher in the years after the drought. The results for singular species evidenced critical conditions for *Castanea sativa* Mill. and *Pinus* species, whereas *Quercus* species showed a progressive decrease in defoliation. Deciduous species, such as *Fagus sylvatica* L., *Ostrya carpinifolia* Scop. and *Quercus pubescens* Willd. suffer the loss of leaves in dry years as a strategy to limit water loss by transpiration but recover their crown in the following years. The recurrence of extreme heat waves and drought from the beginning of the XXI century may increase the vulnerability of forests, and increased tree mortality can be expected in the future.

Keywords: crown conditions; delayed mortality; heat and drought waves; long-term monitoring; ICP Forests; crown recovery

Citation: Bussotti, F.; Papitto, G.; Di Martino, D.; Cocciufa, C.; Cindolo, C.; Cenni, E.; Bettini, D.; Iacopetti, G.; Pollastrini, M. Defoliation, Recovery and Increasing Mortality in Italian Forests: Levels, Patterns and Possible Consequences for Forest Multifunctionality. *Forests* **2021**, *12*, 1476. https://doi.org/10.3390/f12111476

Academic Editor: Mark E. Harmon

Received: 14 September 2021
Accepted: 25 October 2021
Published: 28 October 2021

1. Introduction

Crown defoliation is the most widespread parameter to assess the health and vitality of forest trees in worldwide monitoring programmes [1]. Defoliation is an unspecific parameter, integrating the intrinsic genetic variability of trees, site effects (soil fertility, climatic features, structure and composition of the forest stand) and external factors such as abiotic and biotic stresses [2,3]. The real significance of defoliation and its consequences on physiological functioning and growth, however, is not clear, although several studies have addressed this issue, often reaching contrasting results [4–8]. Defoliation is not necessarily equivalent to physiological damage and can be considered indicative of the dynamic equilibrium of a tree in its own environment, and we assume that the physiological responses may depend not only on the levels of defoliation itself, but also on year-by-year differences and species-specific strategies to cope with stress. The levels and fluctuation of tree defoliation at European levels are published and commented yearly in the ICP Forests Reports [9].

In recent years, attention was devoted to increased tree mortality at the world level, as a complex phenomenon connected to climate change [10,11]. Tree mortality is an essential ecological function that regulates the demographic processes within populations and communities [12,13]. In non-disturbed conditions, tree mortality is a consequence of competition processes and ageing, while occasional forest disturbances, such as fire, windstorms and droughts, and their combination can promote large-scale tree mortality [14]. Tree mortality induces dynamic changes in the structure and composition of forest stands, allowing the maintenance of the diversity and efficiency of forests and assuring in the long period their multifunctionality and provision of ecosystem services.

Climate change is expected to increase the rate of tree mortality and canopy dieback through the recurrence of catastrophic events (extreme heat and drought waves, windstorms, etc.) that weaken trees, making them vulnerable to the attacks of pests or fungal diseases that often represent the ultimate cause of tree death [15,16]. Widespread crown defoliation and mortality affect ecosystem functions and services at different levels: provisioning (loss of forest products), regulating (climate and water regulation) and cultural services (changes in landscape features and perception), as well as a loss of biodiversity and change in specific composition in the understorey and at the soil level.

Mitigation and recovery strategies require measures of proactive silviculture, including a controlled substitution of species according to the principles of assisted migration [17]. In this perspective, it is necessary to know the incidence over time of tree mortality. Tree mortality is assessed with remote sensing and terrestrial surveys, namely national forest inventories and the European transnational monitoring network ICP Forests [11]. Open questions concern the definition of tree mortality, which includes standing dead trees as well uprooted and fallen individuals, but sometimes also trees removed for phytosanitary purposes or planned forest interventions, according to the current definition adopted in the ICP Forests monitoring programme. Conifer trees with all dry leaves can be considered dead, but in many broadleaved species, death is not always a univocal event. Deciduous species may wilt and shed their leaves as a strategy to avoid summer drought, and then later restore their photosynthetic apparatus [18], whereas in other cases they can resprout by dormant buds [19], thus replacing the dead canopy. Another question concerns the temporal repetition of the investigation. Surveys carried out at pluriannual intervals, such as forest inventories, may underestimate the mortality that occurred during the interval between surveys. Yearly surveys on a common sample of trees [20] allow not only the detection of the actual annual rate of tree mortality, but also knowledge of the conditions of trees before death, the causes, as well the possible recovery of partial or completely defoliated (but not dead) trees.

Italian forests have great ecological and climatic variability, including Mediterranean, temperate and alpine regions, each of them with different specific compositions and different sensitivities to climatic and pathological stress. Climate changes and the occurrence of extreme events may have different characteristics in the various climatic and forest regions, but some common traits have been recognised [21–24]. A progressive increase in temperatures was registered in Italy from the 1980s. The overall precipitation on an annual basis showed no significant trends, but the distribution changed with a decrease in the rainy days and an increase in the intensity of precipitation for each rainy event. Moreover, an increase in springtime aridity and the frequency of extreme climatic events were also observed, including heat and drought waves and windstorms. In recent decades, the most important events were the dry spells in 2003 [25], 2008–2010 [26], 2012 [27,28] and 2017 [29] and the windstorm "Vaia" that affected the alpine regions in 2018 [30]. Many stress factors and their interactions are therefore implied in defoliation and mortality. Defoliation, as assessed in the ICP Forests pan-European programme, is considered a good predictor for tree mortality by [31].

In this paper, we analyse the patterns of defoliation and mortality, and possible causes, in Italian forests. We aimed to verify: (i) the background levels of tree mortality and possible increasing mortality over the years; (ii) the relationships between defoliation and mortality; and (iii) the possible processes of recovery after extreme defoliation events or, in other words, if extreme defoliation triggers an irreversible process leading to mortality. Our goal is to verify the usefulness of long-term surveys for the management of forests aimed at the maintenance of ecosystem services.

2. Materials and Methods

2.1. Dataset

The dataset refers to the Italian Level I network for crown condition assessment, consisting of 253 plots distributed across the country, on a 15 × 18 km grid with a mean of about 5000 trees each year, during the period 1997–2020. Standardised methods and common protocols were used for the field surveys [32–34]. Defoliation was evaluated according to a proportion scale in 5% intervals (0 = no defoliation; 5; 10; 15 . . . 100% = dead tree) by comparing the sampled trees with photographic standards of reference trees for the main forest species (i.e., photoguide method [35,36]). Dead trees (standing, fallen and uprooted trees) were no longer evaluated in the years following their death, so the values presented here are the actual yearly mortality (the death that occurred in the year before the survey). Yearly surveys were carried out during the summer months. Field crews were trained before each yearly assessment for comparability and repeatability of the scores [37,38].

The whole sampled population includes more than 60 species, but 80% of trees fall within 10 main tree species: 6 broadleaved (*Fagus sylvatica* L., *Quercus pubescens* Willd., *Quercus cerris* L., *Castanea sativa* Mill., *Ostrya carpinifolia* Scop., *Quercus ilex* L.) and 4 conifers (*Picea abies* (L.) Karst., *Larix decidua* L., *Pinus sylvestris* L. and *Pinus nigra* Arn.). The mean number of trees for each species is listed in Table 1.

Table 1. General data for the main tree species. N = Number of trees. Percent of defoliated trees beyond the following thresholds: Def > 25% (moderate + severe), Def > 60% (severe), Def > 85% (extreme). Percent of dead trees (Def 100%). Percentages of extreme defoliation and dead trees are also presented cumulatively (Def 85–100%). The ratio Dead/Extr expresses the incidence of mortality in relation to extreme defoliation. The data are the mean per species over the period 1997–2020.

Species	N	Def > 25%	Def > 60%	Def > 85%	Def 100%	Def 85–100%	Dead/Extr
Castanea sativa Mill.	475.08	61.36	12.98	3.52	1.08	4.60	0.31
Fagus sylvatica L.	1099.04	30.21	2.96	0.92	0.42	1.34	0.45
Ostrya carpinifolia Scop.	329.75	34.91	6.07	1.49	0.34	1.83	0.23
Quercus cerris L.	589.96	25.08	1.06	0.22	0.49	0.71	2.18
Quercus ilex L.	220.13	23.27	1.12	0.11	0.08	0.19	0.72
Quercus pubescens Willd.	718.88	51.82	5.08	0.98	0.46	1.44	0.47
Larix decidua L.	348.00	19.93	2.17	0.65	0.59	1.24	0.90
Picea abies (L.) Karst.	544.29	19.39	1.79	0.28	0.50	0.78	1.74
Pinus nigra Arn.	192.88	24.92	2.86	0.33	0.75	1.08	2.24
Pinus sylvestris L.	205.46	34.19	3.42	0.58	1.06	1.64	1.85

Fagus sylvatica and *C. sativa* are (more or less) evenly distributed in all the country, whereas other species have a prevalent regional distribution: *Q. cerris* and *Q. pubescens* in central and southern regions; *O. carpinifolia* in northern and central regions; and *Quercus ilex* in the Mediterranean areas. Among conifers, *Picea abies*, *Larix decidua* and *Pinus sylvestris* are concentrated in the northern, alpine regions (Figure S1), whereas the distribution of *P. nigra* is irregular and depends on afforestation programs with the two subspecies *P. nigra* subsp. *nigra* and *P. nigra* subsp. *laricio*.

2.2. Data Analysis

The analyses of defoliation data were carried out by considering the percent of living trees beyond the following defoliation classes: Def ≤ 25% (no or light defoliation); Def > 25% (moderate); Def > 60% (severe); Def > 85% (extreme). Dead trees (Def = 100%) were considered in a separate class. The results are reported as temporal trends whose significances were calculated according to the Spearman coefficient of correlation (r). Levels of significance (p) are reported for $p < 0.05$ (significant), $p < 0.01$ (very significant) and $p < 0.001$ (highly significant). Ns = not significant.

3. Results and Discussion

3.1. Overall Results

The temporal patterns of defoliated and severely defoliated trees (Def > 25 and Def > 60%, excluding dead trees) are not significant (Figure 1A), although the trees with Def > 25% (33.6% average in the 1997–2020 period) show a slightly decreasing trend ($p < 0.1$). The percent of trees with Def > 60% (4.06% average in the 1997–2020 period) shows a peak in the year 2017 (6.31%), corresponding to the extreme drought and heat wave occurring in this year in Italy [39–41]. There is a positive significant trend concerning the increase in trees with extreme defoliation (Def > 85%) and mortality (Def 100%) (Figure 1B). The average rate of extremely defoliated trees for the whole period was 0.96%, with a peak of 2.06 in 2017; mortality (24-years average: 0.57%) increased in subsequent years (2018: 1.05%; 2019: 1.07%), peaking in 2020 (1.21%). Secondary peaks of mortality in the years 2009–2010 occurred after the heat and drought events in 2008 [26]. No significant correlations were found between crown defoliation (for any defoliation threshold) and mortality during the same year, but there was a highly significant correlation ($r = 0.67$, $p > 0.001$) between mortality and the rate of extremely defoliated (but not dead) trees the year before (Figure 2).

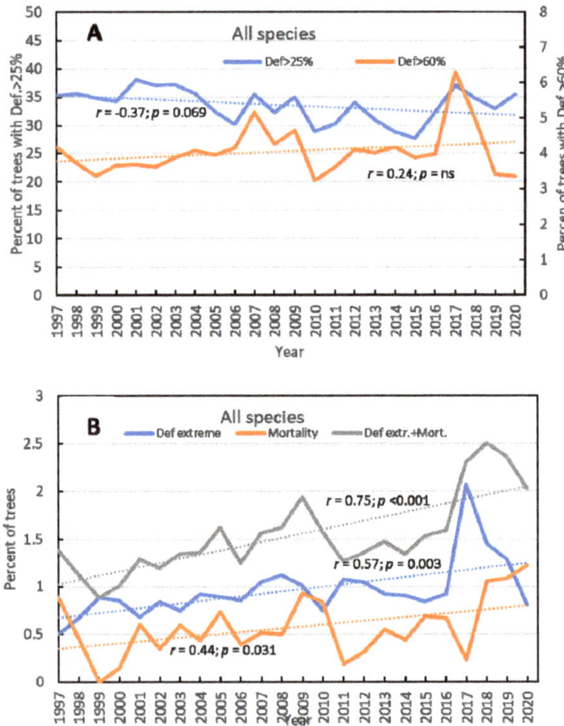

Figure 1. Defoliation and mortality trends (correlation between year and defoliation) over the 1997–2020 period (all species pooled). (**A**) Annual rate of trees with defoliation >25%, and trees with defoliation >60. (**B**) Annual rate of extremely defoliated trees, with defoliation >85%, and dead trees.

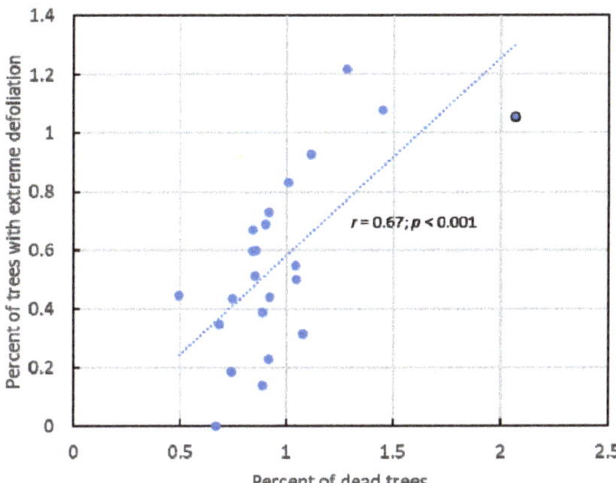

Figure 2. Correlation between the percent of the dead trees assessed during a yearly survey (year n) and percent of extremely defoliated trees the year before (year $n - 1$). All species pooled.

The average levels of mortality are comparable with those reported in the literature. Bertini et al. [42] found, in Italian NFI (National Forest Inventories), that mortality rates by number of trees and by volume (m^3 ha^{-1} y^{-1}) amounted on average to 1.35% and 0.51%, respectively. Neumann et al. [43], analysing the data from the ICP Forests Level I network, found that the mean yearly mortality rate over Europe for the period 2000–2012 was 0.50% with differences among eco-regions, ranging from 0.31% in Central-Western Europe to 1.39% in South-Western Europe. In Switzerland, Etzold et al. [44] analysed over one century of data from forest inventories and found that the long-term average annual mortality rate was 1.5%, with differences between species, tree size and ecoregions, with no significant trends to increase in mortality except for *Pinus sylvestris* L. at low elevations.

Recurrent drought and heat waves are responsible for widespread tree mortality across biomes and climatic regions in the world [15,45,46]. Senf et al. [47], analysing high-resolution annual satellite-based canopy mortality maps across continental Europe from 1987 to 2016, observed that forest mortality was significantly related to drought, and concluded that, overall, drought caused approximately 500,000 ha of excess forest mortality in the same period. Regional impacts leading to crown dieback and mortality were described both for Mediterranean Europe [40,48,49] and Central European countries [50,51]. Delayed tree mortality, as observed in our study, is a phenomenon described in relation to fire and drought impacts [52,53], and it is supposed to be related to carbon starvation as a consequence of the altered pattern on carbon allocation [54].

The conditions of trees in the years before their death can be assessed on the ICP Forests network. The levels of defoliation of trees in the 3 years before death are shown in Figure 3. Two groups of trees are recognisable: one group (A) shows low defoliation until a sudden death; the second group's (B) levels of defoliation increase progressively. These behaviours can be connected to two distinct mortality patterns: death may be a consequence of specific sudden disturbances (for example, insect attacks, windstorms, etc.) on low-defoliated trees in group A or, alternatively, it may indicate a progressive decline until death (high defoliation before death) in group B. Trees with extreme defoliation (Def. > 85%) died in 38% of cases after one year and 70% after four years (Figure 4). However, a portion of trees (4 and 17%) showed light (<30%) and moderate (<65%) defoliation, respectively, after 4 years.

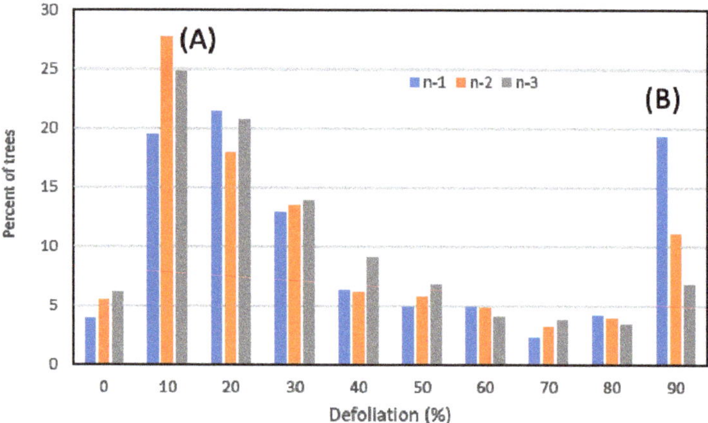

Figure 3. Levels of defoliation of the trees in the three years before a tree death (n = death year; $n-1$, $n-2$, $n-3$ =, respectively, 1, 2, 3 years before death). One group (**A**) shows low defoliation until a sudden death; the second group's (**B**) levels of defoliation increase progressively.

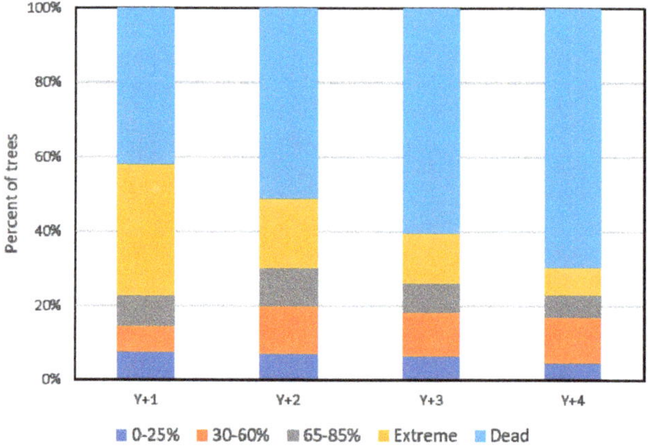

Figure 4. Defoliation rates in the four years after extreme defoliation occurred ($Y+1, \ldots.. Y+4$), according to the usual defoliation classes (0–25%: not or slightly defoliated; 25–60%: moderately defoliated; 60–85%: severely defoliated; 85–100%: extremely defoliated; 100%: dead).

3.2. Results Per Species

The distribution of tree species reflects the different climatic and ecological regions (alpine, temperate, Mediterranean); therefore, defoliation (level and trend) can vary in relation to the intensity of impacts and sensitivity of different species. The levels of defoliation and mortality averaged for the whole assessment period (1997–2020) are shown in Table 1. The patterns of defoliation, i.e., the significance of the trends over the assessment period (1997–2020), are reported in Table 2. The pattern of each species is reported in detail in Supplementary Materials Figure S2.

Table 2. Trends (correlation with year) of tree defoliation and death over the period 1997–2020 (see explication in Table 1).

Species	Def > 25%		Def > 60%		Def > 85%		Def 100%		85 + 100%	
Castanea sativa	0.829	***	0.820	***	0.733	***	0.577	**	0.751	***
Fagus sylvatica	−0.178	ns	0.196	ns	0.314	ns	0.250	ns	0.389	(*)
Ostrya carpinifolia	0.208	ns	−0.101	ns	−0.136	ns	0.422	*	0.043	ns
Quercus cerris	−0.497	*	0.299	ns	0.310	ns	−0.036	ns	0.055	ns
Quercus ilex	−0.680	***	−0.494	*	−0.066	ns	−0.094	ns	−0.093	ns
Quercus pubescens	−0.729	***	−0.468	*	0.366	(*)	−0.029	ns	0.311	ns
Larix decidua	0.208	ns	−0.101	ns	−0.136	ns	0.422	*	0.043	ns
Picea abies	0.240	ns	0.439	*	0.308	ns	0.584	**	0.722	***
Pinus nigra	−0.090	ns	−0.499	*	0.208	ns	0.469	*	0.519	**
Pinus sylvestris	0.417	*	0.524	**	0.224	ns	0.202	ns	0.305	ns

The coefficients of correlation r (Pearson) and p level are shown. * = $p < 0.05$ (significant); ** = $p < 0.01$ (very significant); *** = $p < 0.001$ (highly significant); (*) = $p < 0.1$ (not significant but indicative of a trend). Ns = not significant.

Among broadleaves, *C. sativa* is the most defoliated species for any considered threshold, as well that with the highest mortality rates (1.08% per year). The trends are significantly positive ($p < 0.001$) for each threshold and mortality. High defoliation and mortality rates on *C. sativa* are connected to the phytopathologies affecting this species, including the attack of the Asian wasp *Dryocosmus kuriphilus* Yasumatsu in recent years [55]. In *C. sativa*, the ratio between dead and extremely defoliated trees is quite low.

Oak species show different defoliation levels. *Q. pubescens* shows high defoliation at the Def > 25% and Def > 60% threshold, whereas *Q. cerris* has low defoliation but high mortality. The ratio between dead and extremely defoliated trees is high for *Q. cerris*, indicating that trees with low defoliation levels are also susceptible to death. This behaviour can be related to the so-called "oak decline", a complex phenomenon triggered by drought and fungal infection both in Central and Southern Europe [56] and also described in Italy [57,58]. Low levels of defoliation and mortality have been detected on *Q. ilex*. All the three Mediterranean oak species, both deciduous (*Q. pubescens* and *Q. cerris*) and evergreen (*Q. ilex*), show a negative significant pattern, with decreasing defoliation over time for the thresholds Def > 25% (all oak species) and Def > 60% (*Q. pubescens* and *Q. ilex*).

Fagus sylvatica, the most widespread tree species, shows no significant trends for defoliation levels and mortality. *Ostrya carpinifolia* shows high values of defoliation (especially for the Def > 60% and Def > 85% thresholds) but low levels of mortality. In this latter species, the ratio between dead and extremely defoliated trees is very low (0.23). *Ostrya carpinifolia*, a shallow-rooted species living on steep slopes, sheds its leaves in the driest months of the year as a strategy to withstand drought. This behaviour was described on *O. carpinifolia* by Pollastrini et al. [6] in the forests of Central Italy.

The early loss of leaves was strongly accentuated during the 2017 summer drought, when in *O. carpinifolia* Def > 60% reached the peak of 10.8% and extreme defoliation reached the peak of 3%. *Fagus sylvatica* and *Q. pubescens* increased their levels of severe and extreme defoliation in 2017 (Def > 60%: 7.8% *for F. sylvatica* and 8.9% for *Q. pubescens*; extreme defoliation: 3.6% for *F. sylvatica* and 3.4% for *Q. pubescens*, respectively), with low values of the ratios between dead and extremely defoliated trees. These species recovered partially in the subsequent years (Figure S2).

Among conifers, *P. sylvestris* shows the highest levels of defoliation and mortality, with significant trends over time. The condition of this species reflects its decline in the regions at the southern slopes of the Alps [59–61]. As dry conditions increase, *P. sylvestris* tends to be replaced by *Q. pubescens* [62]. *Pinus nigra* has the highest ratio between dead and extremely defoliated trees, indicating mortality not connected to defoliation. This behaviour may be connected to the ageing and decline of past afforestation programmes [63]. Alpine conifers (*P. abies* and *L. decidua*) have low defoliation levels (Def > 25%). In *P. abies*, there is a significant trend to increasing defoliation (Def > 60%) and mortality, with a high ratio between dead and extremely defoliated trees, probably in relation to frequent biotic [64] and abiotic disturbances (windstorms [30]) causing relevant damage to conifer forests.

3.3. What Have We Learned?

The general conditions of the forests in Italy appear quite stable at the national level across time, despite the recurrent disturbances and extreme climatic events. This apparent stability, however, is the result of different and contrasting species- and site-specific patterns and behaviours, leading to the overall homeostatic capacity of the forests. Whereas some species, such as *C. sativa* and *P. sylvestris*, showed a sharp decline, the conditions of the most important species (*F. sylvatica*, *Q. cerris*, *Q. pubescens*) are stable or even improving according to the usual descriptors of tree health. This stability, however, is threatened by the drought events that in recent years induced an increase in defoliation.

Despite this apparent stability, an increasing fraction of extremely defoliated (Def > 85%) and dead trees is recognisable. Extreme defoliation and mortality are scattered events that may interest singular trees or groups of trees in response to a progressive decline or to sudden stress factors, on ecologically fragile sites and critical tree species. Such situations can expand as a consequence of recurring extreme climate events or new exotic invasive pests. Besides the case of *C. sativa*, the most critical conditions are those affecting the pinewoods (the mountain pines, *P. sylvestris* and *P. nigra* in this paper) and the Alpine conifers (*P. abies*).

The highest levels of mortality were recorded in the 2018–2020 period, i.e., delayed with respect to the year of the drought event (2017). Delayed mortality may be a consequence of the depletion of the reserves of carbon (NSC, non-structural carbohydrates), which reduces the ability of plants to cope with environmental conditions [65,66]. The restoration of carbon reserves may take several years [67], during which trees are vulnerable and potentially sensitive to additional stress factors in the near future. The increasing frequency of recurrent drought events [68–71] can trigger irreversible processes of forest decline that, although still localised, can involve an increasing fraction of trees. It is noticeable that in the last year of observation, 2020, the recovery is still partial for many species (Figure S2). Therefore, it is plausible to hypothesise increasing tree mortality in the following years.

It is noticeable that no anomalies in tree mortality were observed after the extremely dry summer in 2003, but mortality increased after the following events of 2008 and, especially, 2017. A singular event of extreme drought, therefore, may not be sufficient to trigger high tree mortality. Magno et al. [39] report that the 2017 event lasted several months, from November 2016 to October 2017. Moreover, the period of recurrence of drought events (each 4–5 years from the beginning of the century [39]) may be too short to allow the complete physiological recovery of the trees.

4. Conclusions

This 24-year study indicates that the increasing risks for the efficiency of forests can be represented by the rate of dead and extremely defoliated trees, whereas defoliation at the thresholds of Def > 25% or Def > 60% can reflect the acclimatisation of trees to the environmental conditions (including the presence of pests). In this view, the lowering levels of defoliation of xeric Mediterranean oak (both deciduous and evergreen) species may also be indicative of progressive acclimation. The current levels of mortality are not far from the background ones as reported from previous studies [42], but their increasing trend suggests more risks in the future. It is therefore necessary to continue and to improve the current monitoring programmes, enhancing their informative potential with the definition and application of more specific physiological indicators [72–74].

Open questions concern the assessment of relevant but localised events such as the impact of the extreme drought on *Q. ilex* and Mediterranean vegetation in Southern Tuscany in 2017 [40], and the assessment of the potential upcoming risks from alien invasive pests. Future studies require a comprehensive analysis of the environmental conditions that can increase the risks for forests, as well an approach that combines the current field-routine surveys with remote sensing and activities based on citizen science.

Supplementary Materials: The following are available online at https://www.mdpi.com/article/10.3390/f12111476/s1, Figure S1: Distribution of monitoring the plots in Italy (conifers, broadleaves, mixed), Figure S2: Patterns of defoliation of individual forest species.

Author Contributions: F.B.: conceptualisation, statistical analyses, writing original draft. E.C.: data curation, assistance in data acquisition in the field work. D.B.: data curation, assistance in data acquisition in the field work. G.I.: statistical analyses. M.P.: writing and editing. G.P.: coordination of the whole project. D.D.M., C.C. (Cristiana Cocciufa), C.C. (Claudia Cindolo): technical coordination of the project. All authors have read and agreed to the published version of the manuscript.

Funding: This research was funded and carried out within SMART4Action LIFE+ project "Sustainable Monitoring and Reporting to Inform Forest and Environmental Awareness and Protection" LIFE13 ENV/IT/000813.

Data Availability Statement: Data are available in the ICP Forests repository.

Acknowledgments: We are grateful to the personnel of Carabinieri Forestale for the collection of field data. Thanks to Giacomo Colle (Effetreseizero srl) for the re-organisation of the Level I-ICP Forests data for the Italian plots.

Conflicts of Interest: The authors declare no conflict of interest.

References

1. Ferretti, M.; Fisher, R. *Forest Monitoring, Methods for Terrestrial Investigations in Europe with an Overview of North America and Asia*; Develop. Environ. Sci. 12 Series; Krupa, S., Ed.; Elsevier: Amsterdam, The Netherlands, 2013; p. 507.
2. Iacopetti, G.; Bussotti, F.; Selvi, F.; Maggino, F.; Pollastrini, M. Forest ecological heterogeneity determines contrasting relationships between crown defoliation and tree diversity. *For. Ecol. Manag.* **2019**, *448*, 321–329. [CrossRef]
3. Toïgo, M.; Nicolas, M.; Jonard, M.; Croisé, L.; Nageleisene, L.M.; Jactel, H. Temporal trends in tree defoliation and response to multiple biotic and abiotic stresses. *For. Ecol. Manag.* **2020**, *477*, 118476. [CrossRef]
4. Gottardini, E.; Cristofolini, F.; Cristofori, A.; Camin, F.; Calderisi, M.; Ferretti, M. Consistent response of crown transparency shoot growth and leaf traits on Norway spruce (*Picea abies* (L.) H. Karst.) trees along an elevation gradient in northern Italy. *Ecol. Ind.* **2016**, *60*, 1041–1044. [CrossRef]
5. Gottardini, E.; Cristofolini, F.; Cristofori, A.; Pollastrini, M.; Camin, F.; Ferretti, M. A multi-proxy approach reveals common and species-specific features associated with tree defoliation in broadleaved species. *For. Ecol. Manag.* **2020**, *467*, 118–151. [CrossRef]
6. Pollastrini, M.; Feducci, M.; Bonal, D.; Fotelli, M.; Gessler, A.; Gossiord, C.; Guyot, V.; Jactel, H.; Nguyen, D.; Radoglou, K.; et al. Physiological significance of forest tree defoliation: Results from a survey in a mixed forest in Tuscany (central Italy). *For. Ecol. Manag.* **2016**, *361*, 170–178. [CrossRef]
7. Tallieu, C.; Badeau., V.; Allard., D.; Nageleisen, L.M.; Bréda., N. Year-to-year crown condition poorly contributes to ring width variations of beech trees in French ICP level I network. *For. Ecol. Manag.* **2020**, *465*, 118071. [CrossRef]
8. Ferretti, M.; Bacaro, G.; Brunialti, G.; Calderisi, M.; Croisé, L.; Frati, L.; Nicolas, M. Tree canopy defoliation can reveal growth decline in mid-latitude temperate forests. *Ecol. Ind.* **2021**, *127*, 107749. [CrossRef]
9. Michel, A.; Prescher, A.K.; Schwärzel, K. Forest Condition in Europe: The 2020 Assessment. In *ICP Forests Technical Report under the UNECE Convention on Long-Range Transboundary Air Pollution (Air Convention)*; Thünen Institute: Eberswalde, Germany, 2020.
10. Hartmann, H.; Moura, C.F.; Anderegg, W.R.L.; Ruehr, N.K.; Salmon, Y.; Allen, C.D.; Arndt, S.K.; Breshears, D.D.; Davi, H.; Galbraith, D.; et al. Research frontiers for improving our understanding of drought-induced tree and forest mortality. *New Phytol.* **2018**, *218*, 15–28. [CrossRef]
11. Hartmann, H.; Schuldt, B.; Sanders, T.G.M.; Macinnis-Ng, C.; Boehmer, H.J.; Allen, C.D.; Bolte, A.; Crowther, T.W.; Hansen, M.C.; Medlyn, B.E.; et al. Monitoring global tree mortality patterns and trends. Report from the VW symposium 'Crossing scales and disciplines to identify global trends of tree mortality as indicators of forest health'. *New Phytol.* **2018**, *217*, 984–987. [CrossRef]
12. Franklin, J.F.; Shugart, H.H.; Harmon, M.E. Tree death as an ecological process. *BioScience* **1987**, *37*, 550–556. [CrossRef]
13. Harmon, M.E.; Bell, D.M. Mortality in Forested Ecosystems: Suggested Conceptual Advances. *Forests* **2020**, *11*, 572. [CrossRef]
14. Anderegg, W.R.L.; Kane, J.M.; Anderegg, L.D.L. Consequences of widespread tree mortality triggered by drought and temperature stress. *Nat. Clim. Chang.* **2013**, *3*, 30–36. [CrossRef]
15. Anderegg, W.R.L.; Hicke, J.A.; Fisher, R.A.; Allen, C.D.; Aukema, J.; Bentz, B.; Hood, S.; Lichstein, J.W.; Macalady, A.K.; McDowell, N.; et al. Tree mortality from drought, insects, and their interactions in a changing climate. *New Phytol.* **2015**, *208*, 674–683. [CrossRef]
16. Das, A.J.; Stephenson, N.L.; Davis, K.P. Why do trees die? Characterizing the drivers of background tree mortality. *Ecology* **2016**, *97*, 2616–2627. [CrossRef] [PubMed]
17. Bussotti, F.; Pollastrini, M.; Holland, V.; Brüggemann, W. Functional traits and adaptive capacity of European forests to climate change. *Environ. Exp. Bot.* **2015**, *111*, 91–113. [CrossRef]

18. Rohner, B.; Kumar, S.; Liechti, K.; Gessler, A.; Ferretti, M. Tree vitality indicators revealed a rapid response of beech forests to the 2018 drought. *Ecol. Ind.* **2020**, *120*, 106903. [CrossRef]
19. Zeppel, M.J.B.; Harrison, S.P.; Adams, H.D.; Kelley, D.I.; Li, G.; Tissue, D.T.; Dawson, T.E.; Fensham, R.; Medlyn, B.E.; Palmer, A.; et al. Drought and resprouting plants. *New Phytol.* **2015**, *206*, 583–589. [CrossRef]
20. ICP-Forests. 30 Years of Monitoring the Effects of Long Range Transboudary Air Pollution on Forests in Europe and beyond. 2016. Available online: http://icp-forests.net/ (accessed on 10 July 2021).
21. Brunetti, M.; Buffoni, L.; Mangianti, F.; Maugeri, M.; Nanni, T. Temperature, precipitation and extreme events during the last century in Italy. *Glob. Plan. Chang.* **2004**, *40*, 141–149. [CrossRef]
22. Brunetti, M.; Maugeri, M.; Monti, F.; Nanni, T. Temperature and precipitation variability in Italy in the last two centuries from homogenised instrumental time series. *Int. J. Clim.* **2006**, *26*, 345–381. [CrossRef]
23. Fratianni, S.; Acquaotta, F. The Climate of Italy. In *Landscapes and Landforms of Italy*; Soldati, M., Marchetti, M., Eds.; Springer: Cham, Switzerland; pp. 29–38.
24. Bartolini, G.; Betti, G.; Gozzini, B.; Iannuccilli, M.; Magno, R.; Messeri, G.; Spolverini, N.; Torrigiani, T.; Vallorani, R.; Grifoni, D. Spatial and temporal changes in dry spells in a Mediterranean area: Tuscany (central Italy), 1955–2017. *Int. J. Clim.* **2021**, *1*, 1–22.
25. Breda, N.; Huc, R.; Granier, A.; Dreyer, E. Temperate forest trees and stands under severe drought: A review of ecophysiological responses, adaptation processes and long-term consequences. *Ann. For. Sci.* **2006**, *63*, 625–644. [CrossRef]
26. Fontana, G.; Toreti, A.; Ceglar, A.; De Sanctis, G. Early heat waves over Italy and their impacts on durum wheat yields. *Nat. Hazards Earth Syst. Sci.* **2015**, *15*, 1631–1637. [CrossRef]
27. Lorenzini, G.; Pellegrini, E.; Campanella, A.; Nali, C. It's not just the heat and the drought: The role of ozone air pollution in the 2012 heat wave. *Agrochimica* **2014**, *58*, 40–42.
28. Brilli, L.; Moriondo, M.; Ferrise, R.; Dibari, C.; Bindi, M. Climate change and Mediterranean crops: 2003 and 2012, two possible examples of the near future. *Agrochimica* **2014**, *58*, 20–33.
29. Rita, A.; Camarero, J.J.; Nolè, A.; Borghetti, M.; Brunetti, M.; Pergola, N.; Serio, C.; Vicente-Serrano, S.M.; Tramutoli, V.; Ripullone, F. The impact of drought spells on forests depends on site conditions: The case of 2017 summer heat wave in southern Europe. *Glob. Chang. Biol.* **2020**, *26*, 851–863. [CrossRef] [PubMed]
30. Dalponte, M.; Marzini, S.; Solano-Correa, Y.T.; Tonon, G.; Vescovo, L.; Gianelle, D. Mapping Forest windthrows using high spatial resolution multispectral satellite images. *Int. J. Appl. Earth Obs. Geoinf.* **2020**, *93*, 102206. [CrossRef]
31. Dobbertin, M.; Brang, P. Crown defoliation improves tree mortality models. *For. Ecol. Manag.* **2001**, *141*, 271–284. [CrossRef]
32. Eichhorn, J.; Roskams, P.; Potocic, N.; Timermann, V.; Ferretti, M.; Mues, V.; Szepesi, A.; Durrant, D.; Seletkovic, I.; Schröck, H.W.; et al. Part IV: Visual Assessment of Crown Condition and Damaging Agents. In *Manual on Methods and Criteria for Harmonized Sampling, Assessment, Monitoring and Analysis of the Effects of Air Pollution on Forests*; UNECE ICP Forests Programme Coordinating Centre, Ed.; Thünen Institute of Forest Ecosystems: Eberswalde, Germany, 2016; p. 54; ISBN 978-3-86576-162-0. Available online: http://www.icp-forests.org/Manual.htm (accessed on 10 July 2021).
33. Bussotti, F.; Bettini, D.; Cenni, E.; Ferretti, M.; Sarti, C.; Nibbi, R.; Capretti, P.; Stergulc, F.; Tiberi, R. PARTE 2—Valutazione della condizione delle chiome. In *Procedure di Rilievo Nelle Aree di Saggio e Valutazione della Condizione delle Chiome. Manuale di Campagna*; Ministero delle Politiche Agricole, Alimentari e Forestali: Roma, Italy, 2016.
34. Gasparini, P.; Di Cosmo, L.; Rizzo, M. PARTE 1—Procedure di rilievo nelle aree di saggio di Livello I. In *Procedure di Rilievo nelle Aree di Saggio e Valutazione della Condizione delle Chiome. Manuale di Campagna*; Ministero delle Politiche Agricole, Alimentari e Forestali: Roma, Italy, 2016.
35. Müller, E.; Stierlin, H.R. Tree Crown Photos. In *Sanasilva*; Swiss Federal Institute for Forest Snow and Landscape Research: Birmensdorf, Switzerland, 1990.
36. Ferretti, M. (Ed.) *Mediterranean Forest Trees. IA Guide for Crown Assessment*; CEC-UN/ECE: Brussels, Belgium; Geneva, Switzerland, 1994.
37. Ferretti, M.; Bussotti, F.; Cenni, E.; Cozzi, A. Implementation of Quality Assurance procedures in the Italian programs of forest condition monitoring. *Water Air Soil Poll.* **1999**, *116*, 371–376. [CrossRef]
38. Bussotti, F.; Cozzi, A.; Cenni, E.; Bettini, D.; Sarti, C.; Ferretti, M. Measurement errors in monitoring tree crown conditions. *J. Environ. Monit.* **2009**, *11*, 769–773. [CrossRef]
39. Magno, R.; De Filippis, T.; Di Giuseppe, E.; Pasqui, M.; Rocchi, L.; Gozzini, B. Semi-Automatic Operational Service for Drought Monitoring and Forecasting in the Tuscany Region. *Geosciences* **2018**, *8*, 49. [CrossRef]
40. Pollastrini, M.; Puletti, N.; Selvi, F.; Iacopetti, G.; Bussotti, F. Widespread crown defoliation after a drought and heat wave in the forests of Tuscany (central Italy) and their recovery—A case study from summer 2017. *Front. For. Glob. Chan.* **2019**, *2*, 74. [CrossRef]
41. Puletti, N.; Mattioli, W.; Bussotti, F.; Pollastrini, M. Monitoring the effects of extreme drought events on forest health by Sentinel-2 imagery. *J. Appl. Remot. Sen.* **2019**, *13*, 020501. [CrossRef]
42. Bertini, G.; Ferretti, F.; Fabbio, G.; Raddi, S.; Magnani, F. Quantifying tree and volume mortality in Italian forests. *For. Ecol. Manag.* **2019**, *444*, 42–49. [CrossRef]
43. Neumann, M.; Mues, V.; Moreno, A.; Hasenauer, H.; Seidl, R. Climate variability drives recent tree mortality in Europe. *Glob. Chang. Biol.* **2017**, *23*, 4788–4797. [CrossRef]

44. Etzold, S.; Ziemińska, K.; Rohner, B.; Bottero, A.; Bose, A.K.; Ruehr, N.K.; Zingg, A.; Rigling, A. One Century of Forest Monitoring Data in Switzerland Reveals Species and Site-Specific Trends of Climate-Induced Tree Mortality. *Front. Plant Sci.* **2019**, *10*, 307. [CrossRef]
45. Allen, C.D.; Macalady, A.K.; Chenchouni, H.; Bachelet, D.; McDowell, N.; Vennetier, M.; Kitzberger, T.; Rigling, A.; Breshears, D.D.; Hogg, E.H.; et al. A global overview of drought and heat-induced tree mortality reveals emerging climate change risks for forests. *Forest Ecol. Manag.* **2010**, *259*, 660–684. [CrossRef]
46. Taccoen, A.; Piedallu, C.; Seynave, I.; Perez, V.; Gégout-Petit, A.; Nageleisen, L.M.; Bontemps, J.D.; Gégout, J.C. Background mortality drivers of European tree species: Climate change matters. *Proc. R. Soc.* **2019**, *286*, 20190386. [CrossRef]
47. Senf, C.; Pflugmacher, D.; Yang, Z.; Sebald, J.; Knorn, J.; Neumann, M.; Hostert, P.; Seidl, R. Canopy mortality has doubled in Europe's temperate forests over the last three decades. *Nat. Commun.* **2018**, *9*, 4978. [CrossRef] [PubMed]
48. Carnicer, J.; Coll, M.; Ninyerola, M.; Pons, X.; Sánchez, G.; Peñuelas, J. Widespread crown condition decline, food web disruption, and amplified tree mortality with increased climate change-type drought. *Proc. Natl. Acad. Sci. USA* **2011**, *108*, 1474–1478. [CrossRef]
49. Ruiz-Benito, P.; Lines, E.R.; Gómez-Aparicio, L.; Zavala, M.A.; Coomes, D.A. Patterns and drivers of tree mortality in Iberian forests: Climatic effects are modified by competition. *PLoS ONE* **2013**, *8*, e56843. [CrossRef]
50. Schuldt, B.; Buras, A.; Arend, M.; Vitasse, Y.; Beierkuhnlein, C.; Damm, A.; Gharun, M.; Grams, T.E.E.; Hauck, M.; Hajek, P.; et al. A first assessment of the impact of the extreme 2018 summer drought on Central European forests. *Basic Appl. Ecol.* **2020**, *45*, 84–103. [CrossRef]
51. Brun, P.; Psomas, A.; Ginzler, C.; Thuiller, W.; Zappa, M.; Zimmermann, N.E. Large-scale early-wilting response of Central European forests to the 2018 extreme drought. *Glob. Chang. Biol.* **2020**, *26*, 7021–7035. [CrossRef] [PubMed]
52. Bigler, C.; Gavin, D.G.; Gunning, C.; Veblen, T.T. Drought induces lagged tree mortality in a subalpine forest in the Rocky Mountains. *Oikos* **2007**, *116*, 1983–1994. [CrossRef]
53. Furniss, T.J.; Larson, A.J.; Kane, V.R.; Lutz, J.A. Wildfire and drought moderate the spatial elements of tree mortality. *Ecosphere* **2020**, *11*, e03214. [CrossRef]
54. Trugman, A.T.; Detto, M.; Bartlett, M.K.; Medvigy, D.; Anderegg, W.R.L.; Schwalm, C.; Schaffer, B.; Pacala, S.W. Tree carbon allocation explains forest drought-kill and recovery patterns. *Ecol. Lett.* **2018**, *21*, 1552–1560. [CrossRef]
55. Battisti, A.; Benvegnù, I.; Colombari, F.; Haack, R.A. Invasion by the chestnut gall wasp in Italy causes significant yield loss in Castanea sativa nut production. *Agric. For. Entomol.* **2014**, *16*, 75–79. [CrossRef]
56. Thomas, F.M.; Blank, R.; Hartmann, G. Abiotic and biotic factors and their interactions as cause of oak decline in Central Europe. *Forest Path.* **2002**, *32*, 277–307. [CrossRef]
57. Gentilesca, T.; Camarero, J.J.; Colangelo, M.; Nolè, A.; Ripullone, F. Drought-induced oak decline in the western Mediterranean region: An overview on current evidences, mechanisms and management options to improve forest resilience. *iFor. Biogeosci. For.* **2017**, *10*, 796–806. [CrossRef]
58. Colangelo, M.; Camarero, J.J.; Borghetti, M.; Gentilesca, T.; Oliva, J.; Redondo, M.A.; Ripullone, F. Drought and Phytophthora Are Associated With the Decline of Oak Species in Southern Italy. *Front. Plant Sci.* **2018**, *9*, 1595. [CrossRef]
59. Minerbi, S.; Cescatti, A.; Cherubini, P.; Hellrigl, K.; Markart, G.; Saurer, M.; Mutinelli, G. Scots Pine dieback in the Isarco Valley due to severe drought in the summer of 2003. *Forest Obs.* **2006**, *2/3*, 89–144.
60. Vacchiano, G.; Garbarino, M.; Borgogno, M.; Mondino, E.; Motta, R. Evidences of drought stress as a predisposing factor to Scots pine decline in Valle d'Aosta (Italy). *Eur. J. Forest Res.* **2012**, *131*, 989–1000. [CrossRef]
61. Rigling, A.; Bigler, C.; Eilmann, B.; Feldmeyer-Christe, E.; Gimmi, U.; Ginzler, C.; Graf, U.; Mayer, P.; Vacchiano, G.; Weber, P.; et al. Driving factors of a vegetation shift from Scots pine to pubescent oak in dry Alpine forests. *Glob. Chang. Biol.* **2013**, *19*, 229–240. [CrossRef]
62. Eilmann, B.; Weber, P.; Rigling, A.; Eckstein, D. Growth reactions of *Pinus sylvestris* L. and *Quercus pubescens* Willd. to drought years at a xeric site in Valais, Switzerland. *Dendrochronologia* **2006**, *23*, 121–132. [CrossRef]
63. Cenni, E.; Bussotti, F.; Galeotti, L. The decline of *Pinus nigra* Arn. reforestation stand on a limestone substrate: The role of nutritional factors examined by means of foliar diagnosis. *Ann. Sci. For.* **1998**, *55*, 567–576. [CrossRef]
64. Netherer, S.; Kandasamy, D.; Jirosová, A.; Kalinová, B.; Schebeck, M.; Schlyter, F. Interactions among Norway spruce, the bark beetle *Ips typographus* and its fungal symbionts in times of drought. *J. Pest Sci.* **2021**, *94*, 591–614. [CrossRef]
65. Hartmann, H.; Trumbore, S. Understanding the roles of nonstructural carbohydrates in forest trees—From what we can measure to what we want to know. *New Phytol.* **2016**, *211*, 386–403. [CrossRef] [PubMed]
66. Wang, Z.; Zhou, Z.; Wang, C. Defoliation-induced tree growth declines are jointly limited by carbon source and sink activities. *Sci. Total Environ.* **2021**, *762*, 143077. [CrossRef]
67. Galiano, L.; Martinez-Vilalta, J.; Sabaté, S.; Lloret, F. Determinants of drought induced effects on crown condition and their relationship with depletion of carbon reserve in a Mediterranean holm oak forest. *Tree Physiol.* **2012**, *32*, 478–489. [CrossRef]
68. Spinoni, J.; Vogt, J.V.; Naumann, G.; Barbosa, P.; Dosio, A. Will drought events become more frequent and severe in Europe? *Int. J. Climatol.* **2018**, *38*, 1718–1736. [CrossRef]
69. Spinoni, J.; Naumann, G.; Vogt, J.V. Pan-European seasonal trends and recent changes of drought frequency and severity. *Glob. Plan. Chang.* **2017**, *148*, 113–130. [CrossRef]

70. Hari, V.; Rakovec, O.; Markonis, Y.; Hanel, M.; Kumar, R. Increased future occurrences of the exceptional 2018–2019 Central European drought under global warming. *Sci. Rep.* **2020**, *10*, 12207. [CrossRef] [PubMed]
71. Moravec, V.; Markonis, Y.; Rakovec, O.; Svoboda, M.; Trnka, M.; Kumara, R.; Hanel, M. Europe under multi-year droughts: How severe was the 2014–2018 drought period? *Environ. Res. Lett.* **2021**, *16*, 034062. [CrossRef]
72. Bussotti, F.; Pollastrini, M. Observing climate change impacts on European forests: What works and what does not in ongoing long-term monitoring networks. *Front. Plant Sci.* **2017**, *8*, 629. [CrossRef] [PubMed]
73. Bussotti, F.; Pollastrini, M. Traditional and novel indicators of climate change impacts on European forest trees. *Forests* **2017**, *8*, 137. [CrossRef]
74. Bussotti, F.; Pollastrini, M. Revisiting the concept of stress in forest trees at the time of global change and issues for stress monitoring. *Plant Stress* **2021**, *2*, 100013. [CrossRef]

MDPI

Article

The Efficiency of Forest Management Investment in Key State-Owned Forest Regions under the Carbon Neutral Target: A Case Study of Heilongjiang Province, China

Shuohua Liu [1], Zhenmin Ding [2], Ying Lin [3] and Shunbo Yao [1,*]

[1] College of Economics and Management, Northwest A&F University, Xianyang 712100, China; ShuohuaLiu@nwafu.edu.cn
[2] College of Economics and Management, Nanjing Forestry University, Nanjing 210037, China; huanglishanren@njfu.edu.cn
[3] School of Economics and Finance, Xi'an Jiaotong University, Xi'an 710061, China; linying@xjtu.edu.cn
* Correspondence: yaoshunbo@nwafu.edu.cn

Abstract: To explore the temporal and spatial evolution of carbon sinks in state-owned forest regions (SOFRs) and the efficiency of increased carbon sinks, this study used panel data from 19 periods in 40 key SOFRs in Heilongjiang Province from 2001 to 2019. Additionally, combined with geographic information system (GIS) and remote sensing (RS) technology, the individual fixed-effect model was used to estimate the number of forest management investment (FMI) lagging periods, and the panel threshold model was used to investigate the differences in the FMI efficiency in various forest regions. From 2001 to 2019, the carbon sink of key SOFRs in Heilongjiang Province showed an upward trend over time, with a growth rate of 20.17%. Spatially, the phenomenon of "increasing as a whole and decreasing in a small area" was found, and the carbon sink of each forest region varied greatly. The standard deviation ellipse of the carbon sink presented a "southeast–northwest" pattern and had "from southeast to northwest" migration characteristics. The FMI amount from 2001 to 2019 showed an upward trend, with a total of CNY 46.745 billion, and varied greatly among forest regions. Additionally, the carbon sink amount in each SOFR affected the FMI efficiency. The threshold of the model was 5,327,211.8707 tons, and the elastic coefficients of the impact of FMI below and above the threshold on the carbon sink were 0.00953 and 0.02175, respectively. The latter's FMI efficiency was 128.23% higher than that of the former. Finally, the increase in FMI to a carbon sink followed the law of diminishing marginal benefits. Therefore, the government should rationally plan the level of FMI in each SOFR to improve the FMI cost-effectiveness and help achieve the goal of "carbon neutrality".

Keywords: SOFRs; FMI; carbon sink; efficiency; GIS; RS; carbon neutrality

Citation: Liu, S.; Ding, Z.; Lin, Y.; Yao, S. The Efficiency of Forest Management Investment in Key State-Owned Forest Regions under the Carbon Neutral Target: A Case Study of Heilongjiang Province, China. *Forests* **2022**, *13*, 609. https://doi.org/10.3390/f13040609

Academic Editors: Elisabetta Salvatori and Giacomo Pallante

Received: 24 February 2022
Accepted: 13 April 2022
Published: 14 April 2022

Publisher's Note: MDPI stays neutral with regard to jurisdictional claims in published maps and institutional affiliations.

1. Introduction

Faced with a series of issues such as climate change, frequent increases in extreme weather, and industrial structural innovation, China has made a serious commitment to "strive to peak carbon dioxide emissions by 2030 and strive to achieve carbon neutrality by 2060". Under the demand of mitigating climate change, carbon sinks have an important ecosystem regulation function [1–3]. Among them, the main ways to increase sinks are bioenergy with carbon capture and storage (BECCS) and increasing the carbon sink capacity of terrestrial vegetation [4]. The former method has higher costs for technological innovation, investment, and maintenance, and its universality and economy have yet to be considered. Nature-based policies are one of the most cost-effective ways to increase carbon sinks and simultaneously have large social and economic benefits [5]. Among all terrestrial vegetation types, forests are the most important carbon pools in terrestrial ecosystems, accounting for 70%–80% of the global amount of carbon storage [6,7]; additionally, forests have the best long-term effects. Therefore, an increase in the forest carbon sink will play a vital role in the process of carbon neutrality.

State-owned forest regions (SOFRs) are experiencing conflicts between ecological and economic benefits in the process of reform, and they are undertaking the major goal of achieving "carbon neutrality" [8]. Therefore, it is necessary to measure the cost-effectiveness of forest management investment (FMI) in various forest regions and identify areas with high efficiency in forest management. Forest management measures mainly increase the carbon storage of forests by improving the forest growth environment, increasing the forest area, and improving the forest stand quality. SOFRs are large in scale, relatively standardized in their management, and have a high FMI, such as forest tending. The main task of SOFRs is to increase forest resources, improve forest quality, give full play to the production potential of SOFRs, and improve the ecological, social, and economic benefits of various forest regions. However, the improvement of forest quality is not only related to forest management measures, but is also affected by regional forest stand quality, natural climate conditions, and forest region size [9,10]. In particular, forest management has scale effects, and large-scale management results are better [11]. Moreover, because forest tending and afforestation measures do not obtain immediate results, there may be a lag effect. Therefore, the effect of FMI on forest quality improvement is uncertain in time and space, and there may be differences in the level of investment efficiency among forest regions. Therefore, identifying high-efficiency investment areas can help alleviate the conflict between the ecological and economic benefits of SOFRs, and is of great significance for policymakers who wish to improve the efficiency of policy implementation [12].

Research on the ecological efficiency of SOFRs mainly adopts the methods of data envelopment analysis (DEA) and slacks-based measure–data envelopment analysis (SBM–DEA) to evaluate the ecological efficiency of key state-owned forestry enterprises [13,14]. However, this research failed to link the FMI with the increase in the carbon sink amount and focused on the expected and undesired output of the overall input to the ecological environment. Most of the research on the measurement of the ecological effects of forest management has focused on measuring the impact of different forest management methods (natural forests, artificial forests) and forest tending measures on soil and tree diameter at breast height [15–17]. Similarly, most studies from the perspective of increasing carbon sinks in forests have focused on small-scale investigations. The research is based on the results of forest resource surveys used to study the impact of single or several common tree species, such as selective cutting, thinning intensity, tending, and other forest management measures, on the forest stock and carbon cycle per unit area [18,19]. Alternatively, carbon storage can be calculated based on the biomass carbon storage conversion coefficient [20,21]. This type of research typically measures and describes the changes before and after the implementation of single or several tree species forest management measures in a small area, ignoring the influence of other interference factors, and it cannot accurately determine whether changes in carbon sinks are related to forest management measures. Furthermore, there is no correspondence between forest management measures and capital investment, and it is impossible to measure the efficiency of FMI and the difference in the investment efficiency of various forest regions.

In addition, studies have carried out annual calculations on the stock of different tree species to assess the quality of a forest stand. For example, the enhanced vegetation index (EVI) was used to establish a biomass allometric growth and inversion model to measure the changes in the biomass of the different forest species to estimate the changes in the carbon sink amount [22,23]. Furthermore, in the measurement of large-scale forest carbon sinks, some studies have comprehensively measured forest biomass based on land use types and combined with MODIS optical data. The study counted long-term carbon sequestration changes to quantify the carbon sequestration effects of forest management measures such as reforestation, thinning, and forest tending, as well as to assess the related socioeconomic carbon sequestration costs [24–26]. Such studies have comprehensively considered the changes in forest quality and quantity, and there are few research results.

Moreover, most studies on the macro scale use the InVEST model to correspond to the forest species coefficient, and use the change of land type to calculate the total amount of carbon sequestration in forest land [27,28]. However, the accuracy of this calculation method is low. Although the carbon density coefficient is considered, it is mostly empirical and averaged. It can only be used to measure changes in carbon sinks based on changes in land types. In addition, this method does not consider forest quality and its differences, so it is difficult to accurately measure the increase or decrease in the carbon sink amount in the region, and it is impossible to accurately measure the effect of how FMI might increase the carbon sink amount.

In view of this, this paper used panel data from 40 key state-owned forest farms in Heilongjiang Province from 2001 to 2019 to explore the increase in carbon sink efficiency and the differences in FMI in various forest regions under the control of natural meteorological factors. First, this paper combined geographic information system (GIS) and remote sensing (RS) technology based on RS data on vegetation net primary productivity (NPP) and accurately calculated the evolution of carbon sinks in the study area from 2001 to 2019. Second, we controlled for other variables that might affect the increase in forest carbon sinks, introduced the lag term of carbon sinks, used the individual fixed-effect model to determine the FMI lag period, and realized the measurement of the causal relationship between the two. Then, we used the panel threshold model to examine the differences in FMI efficiency under different carbon sink levels and identify forest regions with a higher investment efficiency. Finally, we verified that FMI followed the law of diminishing marginal returns in terms of increasing the carbon sink, and discussed how to rationally plan investment levels. This research aimed to promote the high-quality development of forest resources in SOFRs, increase the level of forest carbon sinks, and improve the utilization efficiency of FMI to achieve "carbon neutrality" as soon as possible.

2. Overview of Study Area, Data, and Methods

2.1. Study Area

Heilongjiang Province is the northernmost and easternmost provincial administrative region of China. It lies between 121°11′–135°05′ E and 43°26′–53°33′ N. The terrain is high in the northwest, north, and southeast and low in the northeast and southwest. It belongs to the cold temperate zone and has a temperate continental monsoon climate. Heilongjiang Province is located along the "Belt and Road" and borders Russia. It is the main body of the terrestrial ecosystem in Northeast Asia. The grassland comprehensive vegetation coverage in Heilongjiang Province is as high as 67.50%, and the forest coverage rate reaches 43.78%, of which key SOFRs account for approximately one-fifth of the province's total area [29]. There are 87 key SOFRs (forest industry enterprises) in northeast and Inner Mongolia, and Heilongjiang Province is in charge of 40. The 40 key state-owned forest divisions are the Yichun Forest Industry Group and Longjiang Forest Industry Group. The Yichun Forest Industry Group has jurisdiction over 17 key SOFRs, and the Longjiang Forest Industry Group has jurisdiction over 23 key SOFRs. There are 627 forest farms (stations) under the jurisdiction of the 40 key SOFRs, and the jurisdictions of four forestry bureaus cross the boundary of Heilongjiang Province [30].

The state-owned forest farm in Heilongjiang Province promotes the development of the forestry economy while promoting the high-quality development of forests, and this approach has had large social and economic benefits. These methods include forest tree breeding and nursery, timber and bamboo harvesting and transportation, economic forest product planting and collection, flowers, wood processing, wood, bamboo, rattan, palm, and reed product manufacturing, forestry tourism and leisure services, and the forest economy. As of 2019, there were nearly 160,000 employees in the industry. In 2019, the total output value reached CNY 39.51412 billion, with a total investment of CNY 6.99164 billion, of which afforestation and forest tending investment reached CNY 2.15375 billion [29]. The geographical location map of the 40 key SOFRs in Heilongjiang Province is shown in Figure 1, where green represents forest land.

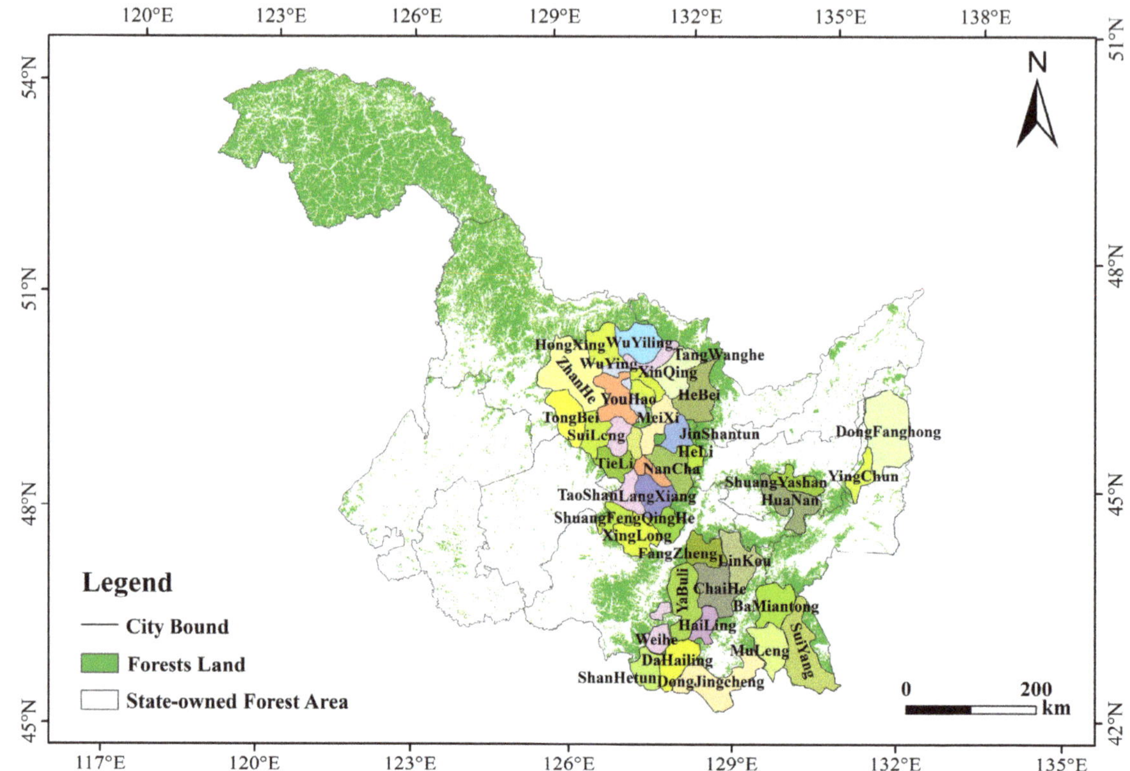

Figure 1. Location map of key SOFRs in Heilongjiang Province.

2.2. Index Selection

(1) Explained variable: As the largest terrestrial carbon pool, forests play a major role in the goal of achieving "carbon neutrality". The management system of China's state-owned forest farms is relatively complete, the FMI is relatively high, and the forest quality is relatively good. Previous studies mostly used NPP as the basis for calculating carbon sinks [31,32]. To measure the carbon sink level of 40 key SOFRs, this paper used vegetation NPP as the basis for calculation. The NPP of vegetation refers to the amount of organic matter accumulated by green plants per unit area and unit time, i.e., the remaining part after deducting autotrophic respiration (RA) from the gross primary productivity (GPP) fixed by plant photosynthesis. This indicator can measure forest quality to the greatest extent and can calculate the carbon sink amount based on basic data. The specific calculation method is explained in the next section.

(2) Explanatory variables: To measure the investment efficiency in forest tending and management and the investment in various major forestry projects, this paper selected the total FMI (invest) as the explanatory variable. This measure includes forest tending investment, forest management and protection investment, forestry fixed asset investment, afforestation and renewal investment, Natural Forest Protection Program (NFPP) investment, and forest quality improvement investment.

(3) Socioeconomic factors: Although the increase in the carbon sink amount requires a large amount of financial investment, it also creates many employment opportunities. Furthermore, as more become engaged in forest management, tending, and industrial development, it has a positive effect on increasing forest quality and increasing the

carbon sink amount. Therefore, the paper incorporated the number of employees (workpop) into the model. In addition, personnel wages play a positive role in stimulating forest management measures such as management and maintenance, and it is necessary to consider the indicator of total wages (wage) [33]. However, forest maintenance measures such as thinning have led to a large amount of timber output and income, which also promotes the development of downstream enterprises related to wood products. Forest maintenance will produce income from tourism, stimulate the economic development of forest regions, and generate income from the tertiary industry. Therefore, in the econometric model, the impact of SOFRs' gross domestic product (gdp) on the carbon sink should be examined [34]. In addition, there may be an inverted U-shaped relationship between the economic development level of the SOFRs and the ecological environment, following the path of the environmental Kuznets curve (EKC) [35]. Therefore, we put gdp and gdp^2 into the model at the same time as control variables.

(4) Natural meteorological factors: The effect and efficiency of FMI are related to forest quality issues. The afforestation land, the growth of young and middle-aged trees, and the effect of forest tending are greatly affected by climatic factors [36,37]. Climatic factors such as precipitation and temperature have greater impacts on forest quality and productivity [38,39]. Additionally, temperature and soil moisture together affect the photosynthesis sensitivity of plants [40,41]. Therefore, when constructing the model, precipitation and temperature need to be used as control variables. Furthermore, because the overall precipitation in the SOFRs of Heilongjiang Province has little difference, the temperature difference between summer and winter is large. Therefore, we used the annual average precipitation and the July average temperature data.

2.3. Data Sources and Processing

(1) Carbon sink data: The carbon sink data were calculated based on the carbon content of vegetation NPP corresponding to vegetation dry matter. The NPP data in this study were derived from the MODIS satellite-based MOD17A3HGF product released by the National Aeronautics and Space Administration (NASA). The spatial resolution of the data is 500 meters, and the time resolution is annual. MRT software and ArcGIS 10.7 software were used to pre-process the data for splicing, cropping, and projection, and the raster calculator function was used to eliminate the NPP data outliers. Finally, the annual NPP sequence data of China from 2001 to 2019 were obtained, and the unit is gC/m^2. After calculating these data, according to the green vegetation photosynthesis chemical formula ($6CO_2 + 6H_2O \rightarrow C_6H_{12}O_6 + 6O_2$), for every 1 kg of dry matter produced by vegetation, 1.63 kg of CO_2 can be fixed, and the carbon content in the dry matter accounts for approximately 45% of the total NPP. Therefore, the calculation formula for the fixed CO_2 of vegetation is as follows: $W_{CO2} = NPP/0.45 \times 1.63$, and the unit of W_{CO2} is g/m^2. Based on this, the carbon sink data for China's forests were calculated in grid form. Then, according to the base map of the 40 key SOFRs, ArcGIS 10.7 was used to extract the carbon sink of each forest region from 2001 to 2019, which was processed according to the unit area to obtain the carbon sink data used in the model.

(2) FMI and socioeconomic data: 1. Data on the FMI, total output value, number of employees, and total wages of each forest region were all sourced from the 2001–2019 China Forestry and Grassland Statistical Yearbook, which is compiled and obtained each year. 2. The administrative boundary data of Heilongjiang Province came from the basic geographic database (http://www.webmap.cn, accessed on 16 July 2021); the vector maps of the 40 key SOFRs were based on the forest base map and drawn according to ArcGIS geographic registration.

(3) Natural data: 1. The basic precipitation data in each forest region came from multiyear station data on the website of the China Meteorological Administration (http://data.cma.cn/, accessed on 20 August 2021). We selected a total of 40 sites in and near

the SOFRs, and we used the ArcGIS 10.7 interpolation analysis function to perform spatial interpolation processing; additionally, kriging was used to interpolate based on the covariance function and to extract the area mean, and the pixel size after processing was 100 m. 2. The average temperature data for each forest region in July came from the Loess Plateau Science Data Center, National Earth System Science Data Sharing Infrastructure, National Science and Technology Infrastructure of China (http://loess.geodata.cn, accessed on 25 August 2021). This dataset is based on the global 0.5° climate dataset released by the CRU (https://crudata.uea.ac.uk/cru/data/hrg/, accessed on 13 July 2021) and the high-resolution climate dataset released by WorldClim (http://www.worldclim.org/, accessed on 6 August 2021). The data were generated by downscaling in China through the Delta space downscaling scheme and had a spatial resolution of approximately 1 km [42]. Then, ArcGIS 10.7 software was used to define the geographic coordinates of the NC data format, project them, and finally extract the annual average temperature in July in the study area. 3. Land use data: The land use data for Heilongjiang Province in 2020 (Figure 1) (1 km × 1 km) came from the Resource and Environmental Science Data Center of the Chinese Academy of Sciences (http://www.resdc.cn/, accessed on 28 July 2021). In this paper, referring to the land type classification standard of the Chinese Academy of Sciences, the study area's land use/cover was reclassified into two categories: forest land and other land. The summary of each variable and the descriptive statistical analysis are shown in Table 1.

Table 1. Variable design and descriptive statistics.

Variable Code	Variable	Unit	Mean	Std. Dev.	Min	Max
carbonsink	Forest carbon sink	Ton	5,357,876.00	3,167,827.00	1,197,632.00	15,400,000
invest	Forest Management Investment	10 thousand	6150.67	6698.97	240.00	31,260.00
gdp	Gross Domestic Product	10 thousand	79,268.55	54,718.58	11,291.84	354,364.00
wage	Total wages of on-the-job employees	10 thousand	7079.68	4884.82	398.60	21,506.10
workpop	Average number of employees	Number of people	5199.82	1986.11	1324.00	14,842.00
pre	Precipitation	mm	639.94	109.03	398.30	934.61
temp	Temperature	°C	20.59	0.96	16.80	23.48

2.4. Research Methods

2.4.1. Analysis of Spatial Distribution Directionality

Spatial distribution directional analysis refers to the outline and dominant direction of the observed variable in the spatial distribution [43]. The standard deviation ellipse (SDE) is a spatial statistical method used to reveal the spatial distribution characteristics of elements [44]. This method mainly measures the centre of gravity, major axis, minor axis, azimuth angle, and other parameters of the SDE of geographic elements to quantitatively describe the spatial distribution characteristics of the observed variables in the study area [45]. The definition formula is as follows:

Center of gravity coordinates:

$$\overline{X}_w = \sum_{i=1}^{n} w_i x_i / \sum_{i=1}^{n} w_i; \overline{Y}_w = \sum_{i=1}^{n} w_i y_i / \sum_{i=1}^{n} w_i \tag{1}$$

$$\tan\theta = \frac{\left(\sum_{i=1}^{n} w_i^2 \widetilde{x}_i^2 - \sum_{i=1}^{n} w_i^2 \widetilde{y}_i^2\right) + \sqrt{\left(\sum_{i=1}^{n} w_i^2 \widetilde{x}_i^2 - \sum_{i=1}^{n} w_i^2 \widetilde{y}_i^2\right)^2 + 4\sum_{i=1}^{n} w_i^2 \widetilde{x}_i^2 \widetilde{y}_i^2}}{2\sum_{i=1}^{n} w_i^2 \widetilde{x}_i \widetilde{y}_i} \tag{2}$$

X-axis standard deviation:

$$\overline{X}_w = \sum_{i=1}^{n} w_i x_i / \sum_{i=1}^{n} w_i; \overline{Y}_w = \sum_{i=1}^{n} w_i y_i / \sum_{i=1}^{n} w_i \tag{3}$$

Y-axis standard deviation:

$$\overline{X}_w = \sum_{i=1}^{n} w_i x_i / \sum_{i=1}^{n} w_i; \overline{Y}_w = \sum_{i=1}^{n} w_i y_i / \sum_{i=1}^{n} w_i \quad (4)$$

In Formulas (1)–(4), $(\overline{X}_w, \overline{Y}_w)$ represents the weighted average centre of each observed variable; (x_i, y_i) represents the spatial coordinates of the observed variable; w_i represents the spatial weight; θ is the azimuth of the standard deviation ellipse, that is, the main trend direction of the data distribution; σ_x, σ_y respectively represent the standard deviation of the ellipse's x-axis and y-axis; and $\widetilde{x}_i$, $\widetilde{y}_i$ respectively represent the coordinate deviation of each observed variable to the weighted average centre.

2.4.2. Individual Fixed-Effects Model

The individual fixed-effects model refers to deterministic variables other than explanatory variables, whose effects vary only with the individual and not with time. Because the carbon sink value of the explained variable in the study was affected by the previous period, the lag term of the explained variable was added to the model explanatory variable. Then, the Hausman test found that the fixed-effects model was better than the random-effects model, so the individual fixed-effects model was used. This paper first built an individual fixed-effects model and judged the regression results of the model without considering the carbon sink as a threshold variable. The model was basically constructed as follows:

$$\begin{aligned} lncarbonsink_{it} = {} & \beta_1 lninvest_{it} + \beta_2 lngdp_{it} + \beta_3 lngdp_{it}^2 + \beta_4 lnwage_{it} + \beta_5 lnworkpop_{it} + \beta_6 llncarbonse_{it} \\ & + \beta_7 lnper_{it} + \beta_8 lntemp_{it} + u_i + \varepsilon_{it} \end{aligned} \quad (5)$$

In Formula (5), i represents the SOFR, and t represents the year. $carbonsink_{it}$ represents the carbon sink of each SOFR; $invest_{it}$ represents the amount of FMI; gdp_{it} represents the gross domestic product of each SOFR; gdp_{it}^2 represents the square of the gross domestic product of each SOFR; $wage_{it}$ represents the total wages; $workpop_{it}$ represents the number of employees in each SOFR; $lcarbonsink_{it}$ represents the carbon sink level after a period of lag; per_{it} represents the annual precipitation; $temp_{it}$ represents the average temperature in July; β_1 to β_8 are the parameters to be estimated in the model; u_i is the individual effect; and ε_{it} is the random disturbance term. If the regression result of the model is unreasonable, the $invest_{it}$ variable was adjusted for the lag period to determine a reasonable lag period.

2.4.3. Panel Threshold Model

The threshold effect means that when a certain parameter reaches a certain critical value (threshold value), it will cause another parameter to change in direction or quantity [46]. The threshold regression model does not need to use cross-terms to determine the nonlinear relationship between FMI and carbon sinks in key SOFRs in Heilongjiang Province, and it was determined by its endogeneity. To determine the specific critical value of the threshold variable, this paper used the panel threshold model and then used the bootstrap method to estimate the significance of the threshold γ. The basic model was set as follows:

$$\begin{aligned} lncarbonsink_{it} = {} & \beta_1 lninvest'_{it} I(carbonse_{it} \leq \gamma) + \beta_2 lninvest'_{it} I(\gamma < carbonse_{it}) + \beta_3 lngdp_{it} + \beta_4 lngdp_{it}^2 \\ & + \beta_5 lnwage_{it} + \beta_6 lnworkpop_{it} + \beta_7 llncarbonse_{it} + \beta_8 lnpre_{it} + \beta_9 lntemp_{it} + \varepsilon_{it} \end{aligned} \quad (6)$$

In Formula (6), γ is the threshold value to be estimated; $I(\cdot)$ is the indicative function, and the value in parentheses is 1 or 0; β_1 to β_9 are the coefficients to be estimated in the model; and other variables indicate the same meaning as those for Formula (5).

2.4.4. Partially Linear Functional-Coefficient Panel Data Model

A partially linear functional-coefficient regression model allows for linearity in some regressors and nonlinearity in other regressors, where the effects of these covariates on the dependent variable vary according to a set of low-dimensional variables nonparametrically [47], thereby showing distinct advantages in capturing nonlinearity and heterogeneity [48]. The main purpose of this model is to avoid the incorrect specification of the function form caused by the linear assumption of the measurement model, and to verify that the effect of the FMI increasing the carbon sink value follows the law of diminishing marginal benefits. Therefore, we set the FMI as a variable with functional coefficients in the function setting, and we set it as variables that enter the functional coefficients that interact with variables in the order specified by the variables that had functional coefficients. The specific form of the model is as follows:

$$carbonsink_{it} = Z'_{it}g(U_{it}) + X'_{it}\beta + \alpha_i + \varepsilon_{it} \quad (7)$$

In Formula (7), i represents the SOFR, and t represents the year. $carbonsink_{it}$ is a scalar dependent variable; $U_{it} = (U_{1,it}, \cdots, U_{l,it})'$ is a vector of continuous variables, that is, $invest_{it}$; $Z_{it} = (Z_{1,it}, \cdots, Z_{l,it})'$ is the vector of covariates in the model; the coefficient is $g(U_{it}) = \{g_1(U_{1,it}), \cdots, g_l(U_{l,it})\}'$; X_{it} is a $k \times 1$ vector of covariates with a constant slope β, which is also $invest_{it}$; α_i is the individual fixed effect that may be related to Z_{it}, U_{it} and X_{it}; and ε_{it} represents the error term.

3. Results and Analysis

3.1. Spatiotemporal Evolution of Carbon Sink and SDE Analysis

3.1.1. Time Change Analysis of Carbon Sink

To investigate the time evolution trend of the carbon sink in the 40 key SOFRs in Heilongjiang Province during the study period, ArcGIS software was used to calculate the annual carbon sink value of each forest region and draw it as a line graph (Figure 2).

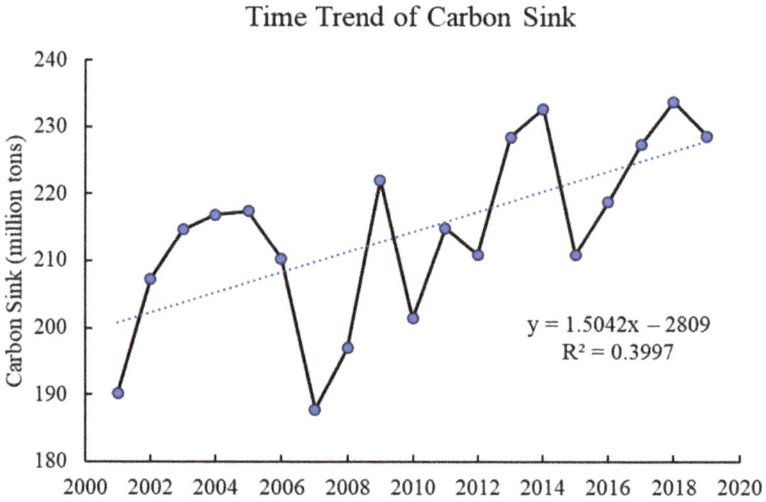

Figure 2. Time change of carbon sinks in the SOFRs in Heilongjiang Province.

From the perspective of time, the carbon sink of the SOFRs showed an upward trend from 2001 to 2019, with a total growth rate of approximately 20.17%, and the fitting formula for annual growth was y = 1.5042x − 2809. It reached the maximum value in 2018, approximately 233.78 million tons, but the overall fluctuation was relatively large. The reasons for the overall increase in the carbon sink are as follows. According to the data from the fifth to eighth forest resource surveys, Heilongjiang has a relatively large proportion of natural forests and a relatively small proportion of artificial forests. The carbon storage per unit area of planted forests is approximately twice that of natural forests. Since 2000, the implementation of the NFPP and the increase in the area of planted forests have increased the intensity of forest management and protection. Therefore, the carbon sink function of forest stands has been enhanced, and the overall trend is increasing.

3.1.2. Spatial Distribution of Carbon Sink and SDE Analysis

To reflect the differences in the carbon sinks of the key SOFRs within the spatial scope, the carbon sink values of the 40 forest regions in 2019 were sorted, and ArcGIS 10.7 was used to link this information with the locations of the forest regions. Then, the natural breaks (Jenks) classification method was selected to classify the carbon sinks in 2019 and visualize them (Figure 3). Additionally, to calculate the direction of the spatial distribution of the carbon sink in the 40 SOFRs, this paper used the spatial statistical tools in ArcGIS 10.7 software to calculate the statistical parameters of the SDEs of the carbon sink in each forest region each year (Table 2).

Regarding the distribution of the carbon sink in each forest region in 2019, the carbon sink of each forest region was quite different, and the overall regional distribution of the carbon sink was characterized by "more in the north and south, but less in the middle", as shown in Figure 3. Among them, the four forest regions of ZhanHe, DongFanghong, DongJingcheng, and SuiYang had the highest carbon sinks, ranging from 9.925 million tons to 14.951 million tons. Mainly due to the large scale of these forest regions and the implementation of key forestry projects, the quality of these forests is better, and thus, the carbon sink value is higher. The seven forest regions of ShuangFeng, DaiLing, HeLi, WuMahe, TangWanghe, WuYing, and ShangGanling had the lowest carbon sinks, with values in the range of 1.683–2.803 million tons. Most likely because of the small scale of these forest regions, the management and maintenance of large-scale forest regions is relatively low.

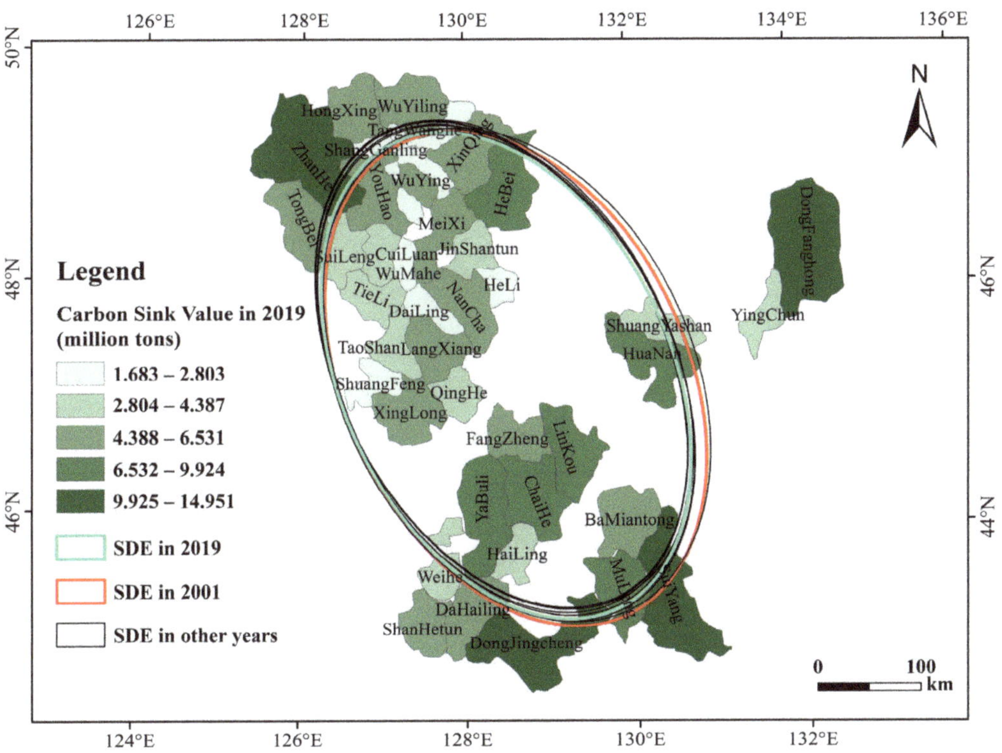

Figure 3. Spatial distribution of the carbon sink and SDE analysis.

Table 2. Changes in the standard deviation ellipse of the carbon sink in the SOFRs.

Year	Shape Length/m	Shape Area/km^2	CenterX/m	CenterY/m	XStdDist/m	YStdDist/m	Rotation/°
2001	1,328,061	130,954.8	1,871,536	5,231,694	164,395.2	253,577.7	155.3266
2003	1,302,777	125,081.2	1,864,519	5,235,355	158,822.4	250,703.2	155.0508
2005	1,307,601	124,978.8	1,860,008	5,240,324	156,842.8	253,660	154.5311
2007	1,337,554	133,949.3	1,872,019	5,235,407	168,557.8	252,970.3	154.9004
2009	1,302,006	123,871.8	1,857,691	5,245,039	156,074.5	252,651.1	154.1664
2011	1,298,014	123,574.6	1,859,176	5,241,175	156,737.9	250,977.7	154.9246
2013	1,297,644	123,315.5	1,858,332	5,245,219	156,223.4	251,276.5	154.0126
2015	1,311,897	126,615.3	1,860,552	5,245,518	159,369.9	252,906.4	153.7399
2017	1,308,664	125,750.8	1,865,522	5,235,737	158,372.2	252,761.9	155.5142
2019	1,297,216	123,603.7	1,862,388	5,234,714	157,096	250,464.6	155.0689

From 2001 to 2019, the SDE of the SOFRs showed an overall pattern of "southeast–northwest" and reflected the characteristics of "from southeast to northwest" migration, but the range of change was small. The turning angle θ showed a fluctuating downward trend, but the magnitude of change was small, indicating that the carbon sink growth rate of the SOFRs in the southeast of the ellipse axis was slightly slower than that of the SOFRs in the northwest of the ellipse axis. The area of the SDE showed a fluctuating downward trend. The overall area of the ellipse decreased by 7351.1 km^2 compared with the value in 2001, and the rate of decrease was 5.61%. The area of the ellipse reached its maximum value in 2007 at 133,949.3 km^2. The short-axis standard deviation fluctuated downwards, decreasing by 7299.2 m; the long-axis standard deviation also fluctuated, decreasing by 3113.1 m. The changes in the carbon sink values in the SOFRs in Heilongjiang Province showed a trend of agglomeration and migration to the northwest.

3.1.3. Analysis of Spatial Changes in the Carbon Sink Value

To investigate the spatial changes in the carbon sink value during the study period, this paper used the spatial distribution maps of the carbon sink (500 m) in the SOFRs in 2001 and 2019. Combined with the function of map algebra in ArcGIS 10.7, the growth rate of the carbon sink at the grid scale was calculated. Then, based on the clustering of the results, the spatial regional carbon sink growth was classified, and the spatial change trend map of the carbon sink in the studied region from 2001 to 2019 was drawn (Figure 4). ArcGIS 10.7 then reclassified the results and calculated the area and corresponding proportions of the five changing trends.

From the perspective of the growth rate of the spatial carbon sink in the 40 key SOFRs in Heilongjiang Province, the carbon sink of the SOFRs from 2001 to 2019 showed an "overall increase and small-scale reduction" phenomenon (Figure 4). Among them, the area where the growth rate of the carbon sink was between 0.151 and 0.300 was the largest, accounting for 36.15%; the second largest growth rate of the carbon sink was between 0.001 to 0.150 and 0.301 to 1.353, accounting for 29.90% and 27.05%, respectively. From 2001 to 2019, the total proportion of positive growth in the carbon sink value in the SOFRs in Heilongjiang Province reached 93.10%. The areas with negative growth rates were divided into two levels, namely, −0.908~−0.086 and −0.087~0.000, and the corresponding area proportions were 2.06% and 4.84%, respectively. Areas with a growth rate of 0.151 or more accounted for 63.20%, indicating that the carbon sink in most regions increased significantly during the study period. Corresponding to the analysis in Figure 3, the carbon sinks of the regions with higher growth rates were lower, which proved that the initial value was lower and the growth potential was greater. However, the regions with growth rates of 0.001 to 0.150 and negative growth rates were mostly located in regions with high carbon sinks, and their growth space was small.

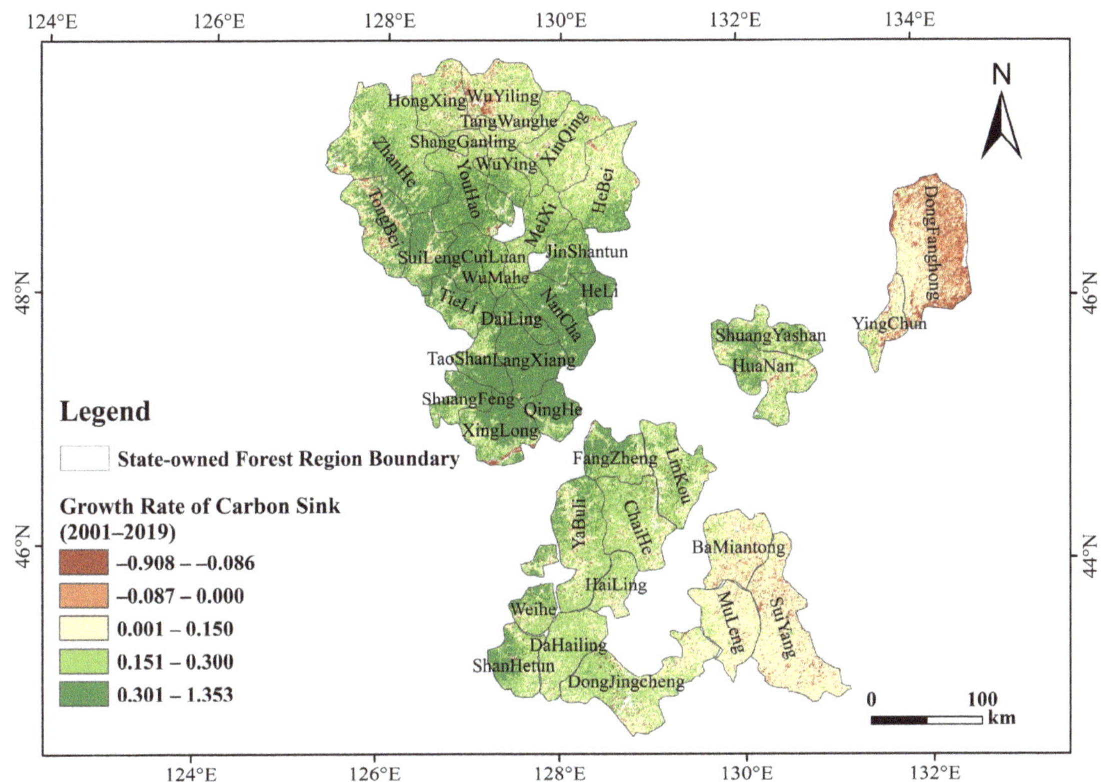

Figure 4. The growth rate of carbon sinks in the key SOFRs in Heilongjiang Province from 2001 to 2019.

The possible reasons for the negative growth of the carbon sink in some regions are as follows. First, the DongFanghong Forest Region has a large scale of forest management and high timber output, which makes forest management and protection more difficult. Second, after 1998, forest management and protection were strengthened, resulting in a decrease in logging, but the geographical distribution was wider. Therefore, there is illegal carbon burning in forest areas with inconvenient transportation and low population density. In addition, it takes a certain period for the area of unforested land to increase as a result of forest tending and other methods to transform into forest land. Therefore, there may be a decrease in the carbon sink value during the study period.

3.2. Analysis of Temporal and Spatial Distribution of Amount of FMI in Key SOFRs

To visually reflect the changes in the annual FMI during the study period, the paper summarized the total annual FMI and created a histogram (Figure 5). To reflect the difference in FMI in the 40 key SOFRs within the spatial scope, the thesis comprehensively organized the total amount of FMI in the 40 SOFRs during the 19-year study period and used ArcGIS 10.7 to connect with the forest regions. The natural break point classification method (Jenks) was chosen to classify the FMI amount and visualize it (Figure 6).

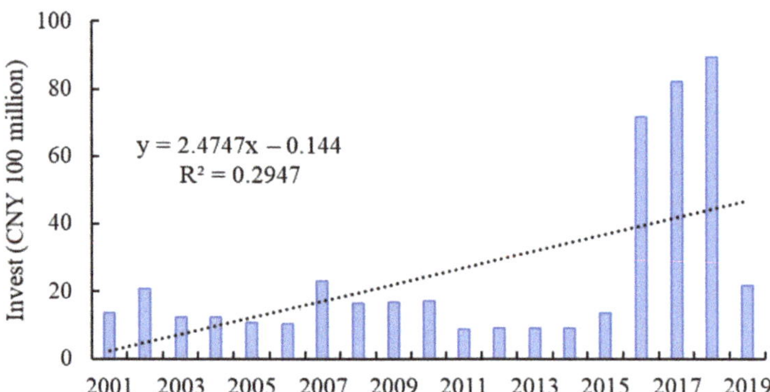

Figure 5. Temporal change in FMI in key SOFRs in Heilongjiang Province.

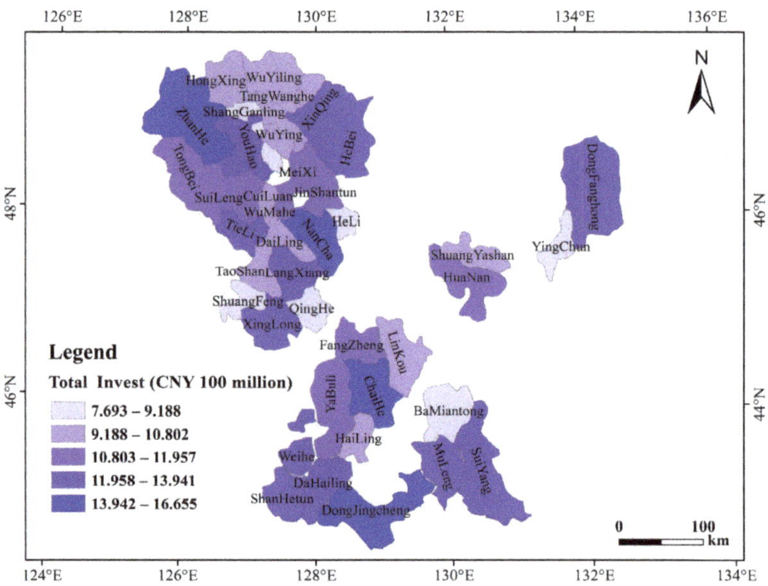

Figure 6. Temporal change in FMI in key SOFRs in Heilongjiang Province.

The temporal change in FMI in the SOFRs showed an overall upward trend. The fitted curve was y = 2.4747x − 0.144, but the R^2 value was low at 0.2947. The investment amount fluctuated from 2001 to 2015. The amount of investment soared from 2016 to 2018 and dropped sharply in 2019. The total FMI from 2001 to 2019 was approximately CNY 46.745 billion. Among them, the highest investment amount was CNY 8.926 billion in 2018. This was followed by that in 2016 and 2017, with values of CNY 7.156 billion and CNY 8.224 billion, respectively. Except from 2016 to 2018, the average FMI in other years was CNY 1.402 billion.

The amount of FMI in various forest regions in Heilongjiang Province was quite different, the regional distribution of FMI was uneven, and there was no obvious aggregation state. Among them, ZhanHe, NanCha, ChaiHe, and DongJingcheng had the highest level of FMI in the four forest regions, ranging from CNY 1.3942 to 1.6655 billion, accounting for 13.55% of the total FMI. Additionally, the amount of FMI in the 12 forest regions of YouHao,

XinQing, HeBei, TieLi, LangXiang, XingLong, DongFanghong, WeiHe, DaHailing, ShanHetun, SuiYang, and MuLeng was in the fourth tier, ranging from CNY 1.195–1.3941 billion, accounting for 33.93%. The amount of FMI in the eight forest regions of TongBei, SuiLeng, JinShantun, CuiLuan, MeiXi, HuaNan, FangZheng, and YaBuli was between CNY 1.0803 and 1.1957 billion, accounting for 19.97%. The investment in the 10 forest regions of Hailing, LinKou, ShuangYashan, TaoShan, DaiLing, WuMahe, HongXing, WuYiling, WuYing, and TangWanghe was between CNY 918.8 million and 1.0802 billion, accounting for 21.79%. In ShangGanling, HeLi, ShuangFeng, QingHe, YingChun, and BaMiantong, the amount of FMI in the six forest regions was among the lowest tiers, ranging from CNY 769.3 to 918.8 million, accounting for only 10.76%.

3.3. Analysis of Effect of Threshold Model Testing

We first examined whether the threshold effect existed, and then determined the number of threshold values based on the test results. According to the analysis of the test results in Table 3, the F statistic and its significance level verified the existence of the threshold effect. The F statistics of the single-threshold model and the dual-threshold model were 78.20 and 90.44, respectively, and both passed the 1% significance level test. The p value indicated that both a single threshold and a double threshold could be selected, and the width of the confidence interval was small. However, a small sample size in the interval would lead to inaccurate estimation results. Therefore, considering the sample size of each interval, the research should select a single threshold model. When the corresponding likelihood ratio statistic LR was 0, the threshold parameter estimated value was γ = 5,327,211.87.

Table 3. Test results of threshold effect.

Threshold Type	F Value	p Value	BS Times	Threshold	Confidence Interval
Single threshold (pfgc1)	78.20 ***	0.0000	300	5,327,211.8707	[5,289,995.8337, 5,341,231.7847]
Double threshold (pfgc1)	90.44 ***	0.0000	300	5,327,211.8707	[5,289,995.8337, 5,341,231.7847]
(pfgc2)				2,494,641.9549	[2,461,170.3222, 2,506,791.0606]

Note: *** indicates significance at the levels of 1%; the p value and critical value are obtained by repeated sampling 300 times using the bootstrap method.

3.4. Analysis of Model Results

The study used Stata 16.1 measurement software to conduct the regression to investigate the efficiency of FMI in increasing carbon sinks in 40 key SOFRs in Heilongjiang Province. The specific results are shown in Table 4.

Model (1) showed the regression result of the individual fixed-effects model, which verified the cost-effectiveness of increasing the carbon sink by FMI without considering the lagging impact of FMI on the carbon sink. Model (2) and Model (3) examined the efficiency of FMI in increasing the carbon sink when FMI lagged by one and two periods. Model (4) used the carbon sink as the threshold variable to examine the differences in the efficiency of FMI under different levels of the carbon sink.

(1) Model (1) showed that the impact of FMI in SOFRs on the carbon sink did not pass the 10% significance test. However, from a theoretical point of view, FMI includes forest tending investment, forest management and protection investment, forestry fixed asset investment, afforestation and renewal investment, NFPP investment, forest quality improvement investment, and other aspects. These funds will increase the forest stock to varying degrees, thereby increasing the forest carbon sink, but the results of the model had no significant impact. The possible reasons are as follows. First, after the investment of forestry funds, the implementation of measures such as afforestation and forest tending take a certain amount of time. Second, some forest management and protection measures do not have an immediate positive impact on

forest carbon sinks. For example, thinning will cause a short-term decline in carbon storage. In addition, measures such as forest tending and artificial afforestation require a certain amount of time to affect the carbon sink, and they do not have an immediate impact on the carbon sink. Therefore, it takes a certain period from the investment to produce obvious effects—that is, the influence of the variable $lninvest_{it}$ on $lncarbonsink_{it}$ lags.

(2) We determined the reasonable lag period for FMI in 40 key SOFRs and used this as a basis to analyze the efficiency of increasing the carbon sink, such as in Models (2) and (3). Since we took the logarithm of all variables, the coefficient in the result indicated the sensitivity of the carbon sink to changes in FMI. The results showed that the elasticity coefficients of FMI for the first and second periods of lag were 0.01419 and 0.00750, respectively. The elasticity coefficient (sensitivity) was the largest when the FMI lagged for a period, which was 0.01419. This result means that for every 1% increase in the amount of FMI, the carbon sink will increase by 0.0142%. Therefore, we chose one lagging period of FMI as a reasonable lag period.

(3) The efficiency of FMI in forest regions with high carbon sinks was greater than that of regions with low carbon sinks, as shown in Model (4). At different carbon sink levels, FMI had a significant positive impact on the carbon sink ($p < 0.01$), but there was strong heterogeneity. The carbon sink of each forest region can, to a large extent, represent the scale of the forest region and the endowment of forest resources. The areas with higher carbon sinks were mostly forest regions with large scales and better endowments of forest resources. Due to the existence of the scale effect, the large-scale forestry management effect was obviously better [11]. When the carbon sink was less than 5,327,211.8707 tons, FMI had a significant positive impact on the carbon sink ($p < 0.01$), and the elasticity coefficient was 0.00953. That is, for every 1% increase in FMI, the carbon sink value of the region increased by 0.00953%. When the carbon sink was higher than 5,327,211.8707 tons, FMI had a significant positive impact on the carbon sink ($p < 0.01$), and its elasticity coefficient was 0.02175. That is, for every 1% increase in FMI, the carbon sink value of the region increased by 0.02175%. The higher the value of the elasticity coefficient of FMI was, the greater the contribution rate of FMI to the growth of the carbon sink, and the higher the investment efficiency. The latter's FMI efficiency was 128.23% higher than that of the former. According to the threshold regression results, Figure 7 was drawn, in which the pink and blue areas are areas with low and high FMI efficiency, respectively.

Among the control variables (Models 1, 2, 3, and 4), the GDP has an inverted U-shaped impact on the carbon sink. That is, as the level of forestry output value in each forest region increased, the carbon sink showed a trend of "rising first, then falling", following the basic path of the environmental Kuznets curve. In Model (4), the number of employees had a significant positive impact on the carbon sink ($p < 0.01$), and the coefficient value was 0.03713. Precipitation had a significant positive impact on the carbon sink ($p < 0.01$), with a coefficient value of 0.22435. That is, for every 1% increase in precipitation, the carbon sink increased by 0.22435%. The results met the tree growth principles and theoretical expectations, thus verifying the logic of the model and the reliability of the results.

(4) The FMI shows a law of diminishing marginal benefits for increasing carbon sinks in each forest region, as shown in Figure 8. With the increase in FMI, the effect of increasing the carbon sink showed a downward sloping curve. The result passed the significance test, and the confidence interval of the coefficient value at each point did not include 0. Since this model was the result of nonparametric estimation, the estimation result of FMI on the carbon sink parameters in the individual fixed-effects model should be an oblique upward curve, and the coefficient value should be positive. When the investment amount in the forest region exceeded CNY 100 million, the growth rate of the carbon sink remained at a low level. However, the marginal benefit

curve did not intersect or approach 0, indicating that FMI has not yet reached the optimal investment scale, and additional investment should be reasonable.

Table 4. Test results of threshold effect.

lncarbonsink	Model (1)		Model (2)		Model (3)		Model (4)	
	Coef.	Std. Err.	Coef.	Std. Err.	Coef.	Std. Err.	Coef.	Std. Err.
lninvest	0.0025	0.0032						
l1lninvest			0.0142 ***	0.0028				
l2lninvest					0.0075 ***	0.0025		
l1lninvest (carbonsink < 5,327,211.87)							0.0095 ***	0.0036
l1lninvest (carbonsink > 5,327,211.87)							0.0218 ***	0.0036
lngdp	−0.4726 ***	0.1355	−0.4226 ***	0.1353	−0.5291 ***	0.1517	−0.3747 ***	0.1178
lngdp2	0.0225 ***	0.0062	0.0205 ***	0.0062	0.0251 ***	0.0069	0.0184 ***	0.0053
lnwage	0.0053	0.0057	0.0027	0.0058	0.0043	0.0059	−0.0048	0.0067
lnworkpop	0.0218 **	0.0108	0.0342 ***	0.0125	0.0302 **	0.0127	0.0371 ***	0.0117
llncarbonsink	0.1633 ***	0.0151	0.1634 ***	0.0143	0.1343 ***	0.0173	0.1551 ***	0.0299
lnpre	0.2460 ***	0.0309	0.2352 ***	0.0301	0.2370 ***	0.0309	0.2244 ***	0.0183
lntemp	−0.0147	0.0517	−0.1008 **	0.0442	0.0096	0.0508	−0.0596	0.0929
_cons	13.5153 ***	0.9451	13.3558 ***	0.9058	14.1423 ***	1.0501	13.1809 ***	0.9350

Note: *** and ** indicate significance at the levels of 1% and 5%, respectively; llncarbonsink is the lncarbonsink value of lagging phase one; l1lninvest is the lninvest value of lagging phase one, and the rest may be deduced by analogy. To reduce the possible heteroscedasticity in the model, we took the logarithm of the variables on both sides of the formula.

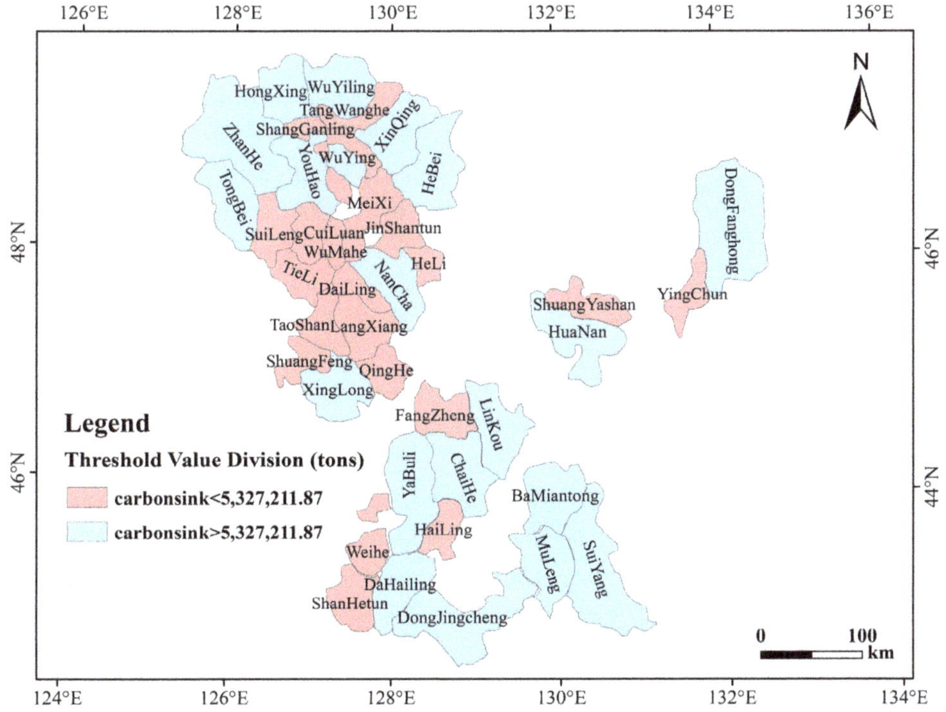

Figure 7. Division of sample units of threshold model.

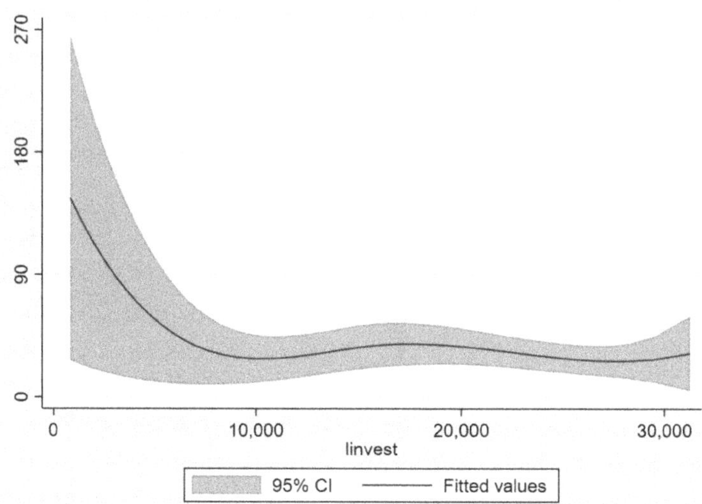

Figure 8. Partially linear functional-coefficient panel data model estimation results.

4. Discussion

This paper used panel data from 40 SOFRs in Heilongjiang Province from 2001 to 2019 for 19 periods, selected the panel threshold model to divide the carbon sink threshold, and examined the efficiency and difference of FMI in increasing the carbon sink value under different carbon sink levels. The analysis of the results showed that the amount of carbon sink greatly affected the efficiency of FMI and verified the law of diminishing marginal returns of FMI. The main differences from previous studies are as follows: (1) We visualized the changes in the carbon sink and FMI in key SOFRs in Heilongjiang Province in time and space to facilitate dynamic analysis. (2) We excluded the sensitivity of FMI changes to carbon sinks on a large scale. (3) We used the panel threshold model to fully investigate the impact of the carbon sink in each forest region on investment efficiency, and then identified efficient FMI areas. (4) Partially linear functional-coefficient panel data models were used to verify that the FMI in the key SOFRs in Heilongjiang Province followed the law of diminishing marginal returns. The research significance of this article mainly lies in the following: carbon sinks were used to characterize the natural conditions, forest resources, and scale of each forest region, and natural factors were included in the system to measure the efficiency of FMI in increasing carbon sinks. According to the differences in FMI in different forest regions, the results provided a decision-making basis for SOFR management, capital allocation, and high-quality forest development, and promoted the realization of the "dual carbon" goal.

During the study period, the overall carbon sink level of the key SOFRs in Heilongjiang Province fluctuated greatly. The possible reason is that the SOFRs, as one of the country's timber supply producing areas, have an annual timber output of up to 4 million cubic meters, and the objects of logging are mostly mature forests and overmature forests. The carbon storage in mature forests and overmature forests is higher than that in young, middle-aged, and near-mature forests, and logging will result in a short-term decline in carbon storage. Additionally, the annual net productivity of young forests and middle-aged forests will reach the highest value after a certain period of time and do not immediately contribute to the increase in carbon storage. In addition, measures such as forest tending and artificial afforestation require a certain amount of time to affect the carbon sink, so immediate results are not visible. In addition, forests are affected by multiple factors, such as pests, wind breaks, and drought. Especially in 2007 and 2011, the annual rainfall was significantly lower than that in other years, which restricted the growth of trees and caused a short-term decline in the carbon sink. In addition, in 2015, the state issued a policy stating that logging was prohibited in all natural forests in northeast China [49,50], and the carbon sinks of forest regions increased. Therefore, there were large fluctuations in the changes in the carbon sink in the study area.

Previous research methods did not consider the impacts of forest scale and resources on the efficiency of FMI. Therefore, it was impossible to accurately identify high-efficiency investment areas, and it was difficult to improve the level of carbon sinks and the efficiency of FMI in a targeted manner, resulting in a waste of forestry funds. In this paper, the investment efficiency of forest management in the key SOFRs of Heilongjiang Province, divided according to the amount of carbon sink, was quite different. There were 21 forest regions with carbon sinks of less than 532,721.87 tons, and their FMI efficiency was 56.18% lower than that of forest regions with more than 532,721.87 tons. In forest regions with low FMI efficiency, first, the amount of FMI should be reasonably controlled, and limited funds should be allocated to forest regions with higher efficiency. Second, the management and supervision of the use of forestry funds should be strengthened, disease and insect pests and drought should be prevented in a timely manner, and timely thinning and replanting should be carried out to improve forest quality in SOFRs.

This paper verified that the effect of increasing the carbon sink of FMI reflected the law of diminishing marginal benefits. According to the results of the model, when the FMI was greater than CNY 100 million, the marginal benefit remained at a stable low level. This result means that when the FMI reached CNY 100 million, the carbon sink volume did not change significantly. However, the marginal benefit curve did not intersect at 0, indicating that FMI has not yet reached the optimal investment scale. Therefore, the government should reasonably increase the amount of FMI when planning FMI in various forest regions. However, since all FMI is used to improve forest quality, the growth potential of each unit area of forest in a certain period of time is limited. Therefore, even without considering the problems in the process of fund management and use, there is a diminishing marginal benefit for each share of funds to increase the carbon sink. Generally, after the forest quality reaches a certain level, even if the FMI gradually increases, the forest quality tends to have a stable and slow growth state. Therefore, the effect of increasing the carbon sink of FMI shows a law of diminishing marginal benefits.

The main contribution and innovation of the research lies in the establishment of an evaluation system for FMI efficiency. In addition, natural factors and forest resource characteristics are incorporated into the system to measure the utilization efficiency. Our research will help to explore the reasons for the differences in FMI efficiency, promote forest management according to local conditions, and improve the utilization efficiency of SOFRs. Furthermore, our research constructs a calculation method for the optimal investment scale for the purpose of ecological benefit. It plays an important role in improving government budget and final accounts, and provides a decision-making basis for forest management, fund allocation, and high-quality forest development. Of course, our research has limitations. Our research area was limited to SOFRs in Heilongjiang Province, and the results of the study have directional policy significance for SOFRs. However, due to regional heterogeneity, the applicability of this result to other SOFRs remains to be further verified. Future research can use the methods described in this paper to explore the difference in the efficiency of state-owned FMI and collective FMI in increasing carbon sinks. In addition, future research can subdivide the efficiency of increasing the carbon sink in various SOFRs and provide targeted suggestions for government FMI management and decision making.

5. Conclusions

This paper used the panel data from 40 SOFRs in Heilongjiang Province from 2001 to 2019 as the research sample. First, the individual fixed-effects model was used to determine a reasonable lag period, and then the panel threshold model was selected to investigate the increase in carbon sink efficiency and differences in FMI under different carbon sink levels. In addition, partially linear functional-coefficient panel data models were used to verify the law of the effect of FMI on increasing the carbon sink. The main conclusions of the study were as follows. (1) The aggregate carbon amount of SOFRs in Heilongjiang Province from 2001 to 2019 showed an overall upward trend, the total growth rate was 20.17%, and the overall fluctuation was large. In terms of space, the carbon sink showed a phenomenon of "increasing as a whole and decreasing in a small area". The SDE of the carbon sink presented a pattern of "southeast–northwest" and showed the characteristics of "from southeast to northwest" migration. In addition, the carbon sink of each forest region was quite different, showing the characteristics of "more in the north and south, but less in the middle". (2) The total FMI from 2001 to 2019 was approximately CNY 46.745 billion, showing an upward trend, but the amount of FMI varied greatly among forest regions. (3) The amount of carbon sink in each forest region significantly affected the efficiency of increasing the carbon sink by FMI. When the carbon sink was less than 5,327,211.8707 tons, the elasticity coefficient of the impact of FMI on the carbon sink was 0.00953. When the carbon sink was higher than 5,327,211.8707 tons, the elasticity coefficient of the impact of FMI on the carbon sink was 0.02175, and the latter's FMI efficiency was 128.23% higher than that of the former. (4) The increasing carbon sink effect of FMI showed the law of diminishing marginal benefits, but it has not yet reached the optimal investment scale. When the FMI reached CNY 100 million, the growth rate of the carbon sink remained at a low level.

Overall, the government should reasonably increase the level of FMI in various forest regions and simultaneously strengthen the management and supervision of the use of forest management funds. Additionally, timely prevention of plant diseases and insect pests, drought prevention and moisture conservation, and timely implementation of forestry measures such as thinning and replanting should be implemented. The high-quality development of forests in the key SOFRs of Heilongjiang Province should be promoted and the realization of the "dual carbon" goal can be achieved.

Author Contributions: Conceptualization, S.L. and S.Y.; methodology, S.L.; software, S.L. and Z.D.; validation, Z.D., Y.L. and S.Y.; formal analysis, S.L. and Z.D.; investigation, S.L., Y.L. and Z.D.; resources, Y.L. and S.Y.; data curation, S.L.; writing—original draft preparation, S.L.; writing—review and editing, S.L.; visualization, S.L., Y.L. and Z.D.; supervision, S.Y. and Z.D.; project administration, S.Y.; funding acquisition, S.Y. All authors have read and agreed to the published version of the manuscript.

Funding: This research was funded by the Special Fund for Scientific Research of Forestry Commonwealth Industry, grant number 201504424 and the National Natural Science Foundation of China, grant number 71473195, 71773091.

Institutional Review Board Statement: Not applicable.

Data Availability Statement: No new data were created or analyzed in this study. Data sharing is not applicable to this article.

Conflicts of Interest: The authors declare no conflict of interest.

References

1. Cho, S.H.; Soh, M.; English, B.C.; Yu, T.E.; Boyer, C.N. Targeting payments for forest carbon sequestration given ecological and economic objectives. *For. Policy. Econ.* **2019**, *100*, 214–226. [CrossRef]
2. Daigneault, A.; Favero, A. Global forest management, carbon sequestration and bioenergy supply under alternative shared socioeconomic pathways. *Land Use Policy* **2021**, *103*, 105302. [CrossRef]
3. Cho, S.H.; Lee, J.; Roberts, R.; Yu, E.T.; Armsworth, P.R. Impact of market conditions on the effectiveness of payments for forest-based carbon sequestration. *For. Policy. Econ.* **2018**, *92*, 33–42. [CrossRef]
4. Weng, Y.; Cai, W.; Wang, C. Evaluating the use of BECCS and afforestation under China's carbon-neutral target for 2060. *Appl. Energy* **2021**, *299*, 117263. [CrossRef]
5. Bithas, K.; Latinopoulos, D. Managing tree-crops for climate mitigation. An economic evaluation trading-off carbon sequestration with market goods. *Sustain. Prod. Consum.* **2021**, *27*, 667–678. [CrossRef]

6. Baccini, A.; Goetz, S.J.; Walker, W.S.; Laporte, N.T.; Sun, M.; Sulla-Menashe, D.; Hackler, J.; Beck, P.S.A.; Dubayah, R.; Friedl, M.A.; et al. Estimated carbon dioxide emissions from tropical deforestation improved by carbon-density maps. *Nat. Clim. Chang.* **2012**, *2*, 182–185. [CrossRef]
7. Mitchard, E.T.A. The tropical forest carbon cycle and climate change. *Nature* **2018**, *559*, 527–534. [CrossRef]
8. Cao, Y.K.; Li, M.Y.; Li, D.X.; Zhu, Z.F. Research on the Reform of Heilongjiang Province State-owned Forest Region and Its Impact on the Natural Forest Protection Project. *For. Econ. Issues* **2020**, *40*, 225–235. (In Chinese) [CrossRef]
9. He, X.; Lei, X.D.; Dong, L.H. How large is the difference in large-scale forest biomass estimations based on new climate-modified stand biomass models? *Ecol. Indic.* **2021**, *126*, 107569. [CrossRef]
10. Kweon, D.; Comeau, P.G. Climate, site conditions, and stand characteristics influence maximum size-density relationships in Korean red pine (*Pinus densiflora*) and Mongolian oak (*Quercus mongolica*) stands, South Korea. *For. Ecol. Manag.* **2021**, *502*, 119727. [CrossRef]
11. Liu, M.X.; Zhu, Z.; Shen, Y.Q.; Li, B.W.; Yu, K. Research on the impact of forestland management scale on the selection and effect of forestry subsidy policy implementation targets: Based on the survey of two counties and cities in Zhejiang Province. *Manag. Rev.* **2021**, *33*, 82–91. (In Chinese) [CrossRef]
12. Ding, Z.M.; Yao, S.B. Ecological effectiveness of payment for ecosystem services to identify incentive priority areas: Sloping land conversion program in China. *Land Use Policy* **2021**, *104*, 105350. [CrossRef]
13. Xue, H.; Frey, G.E.; Geng, Y.D.; Cubbage, F.W.; Zhang, Z.H. Reform and efficiency of state-owned forest enterprises in Northeast China as "social firms". *J. For. Econ.* **2018**, *32*, 18–33. [CrossRef]
14. Ning, Y.L.; Liu, Z.; Ning, Z.K.; Zhang, H. Measuring eco-efficiency of state-owned forestry enterprises in Northeast China. *Forests* **2018**, *9*, 455. [CrossRef]
15. Gong, C.; Tan, Q.Y.; Liu, G.B.; Xu, M.X. Forest thinning increases soil carbon stocks in China. *For. Ecol. Manag.* **2021**, *482*, 118812. [CrossRef]
16. Feng, Q.Y.; Chen, C.F.; Qin, L.; He, Y.T.; Wang, P.; Duan, Y.X.; Wang, Y.F.; He, Y.J. The influence of different management modes on the stand structure and plant diversity of *Quercus mongolica* natural secondary forest. *For. Sci.* **2018**, *54*, 12–21. (In Chinese) [CrossRef]
17. Zhang, H.N.; Chen, S.F.; Xia, X.; Ge, X.M.; Zhou, D.Q.; Wang, Z. The competitive mechanism between post-abandonment Chinese fir plantations and rehabilitated natural secondary forest species under an in situ conservation policy. *For. Ecol. Manag.* **2021**, *502*, 119725. [CrossRef]
18. Bravo-Oviedo, A.; Ruiz-Peinado, R.; Modrego, P.; Alonso, R.; Montero, G. Forest thinning impact on carbon stock and soil condition in Southern European populations of *P. sylvestris* L. *For. Ecol. Manag.* **2015**, *357*, 259–267. [CrossRef]
19. Kim, C.; Son, Y.; Lee, W.K.; Jeong, J.; Noh, N.J. Influences of forest tending works on carbon distribution and cycling in a Pinus densiflora S. et Z. stand in Korea. *For. Ecol. Manag.* **2009**, *257*, 1420–1426. [CrossRef]
20. Zhang, C.H.; Wang, L.Y.; Song, Q.W.; Chen, X.F.; Gao, H.; Wang, X.Q. Forest carbon storage and its dynamic changes in Heilongjiang Province from 1973 to 2013. *China Environ. Sci.* **2018**, *38*, 4678–4686. (In Chinese) [CrossRef]
21. Ahmad, B.; Wang, Y.H.; Hao, J.; Liu, Y.H.; Bohnett, E.; Zhang, K.B. Variation of carbon density components with overstory structure of larch plantations in northwest China and its implication for optimal forest management. *For. Ecol. Manag.* **2021**, *496*, 119399. [CrossRef]
22. Wu, J.C.; Zhou, J.; Zhang, Y.; Yu, X.Y.; Shi, L.; Qi, L.H. The dynamic changes of the value of moso bamboo forests to sequester carbon and increase sink: Taking Fujian Province as an example. *For. Sci.* **2020**, *56*, 181–187. (In Chinese) [CrossRef]
23. Chen, D.S.; Huang, X.Z.; Zhang, S.G.; Sun, X.M. Biomass modeling of larch (*Larix* spp.) plantations in China based on the mixed model, dummy variable model, and Bayesian hierarchical model. *Forests* **2017**, *8*, 268. [CrossRef]
24. Tong, X.; Brandt, M.; Yue, Y.; Ciais, P.; Rudbeck Jepsen, M.; Penuelas, J.; Wigneron, J.P.; Xiao, X.; Song, X.P.; Horion, S.; et al. Forest management in southern China generates short term extensive carbon sequestration. *Nat. Commun.* **2020**, *11*, 1–10. [CrossRef]
25. Austin, K.G.; Baker, J.S.; Sohngen, B.L.; Wade, C.M.; Daigneault, A.; Ohrel, S.B.; Ragnauth, S.; Bean, A. The economic costs of planting, preserving, and managing the world's forests to mitigate climate change. *Nat. Commun.* **2020**, *11*, 1–9. [CrossRef]
26. Lefebvre, D.; Williams, A.G.; Kirk, G.J.D.; Paul, B.J.; Meersmans, J.; Silman, M.R.; Román-Dañobeytia, F.; Farfan, J.; Smith, P. Assessing the carbon capture potential of a reforestation project. *Sci. Rep.* **2021**, *11*, 2–11. [CrossRef]
27. Babbar, D.; Areendran, G.; Sahana, M.; Sarma, K.; Raj, K.; Sivadas, A. Assessment and prediction of carbon sequestration using Markov chain and InVEST model in Sariska Tiger Reserve, India. *J. Clean. Prod.* **2021**, *278*, 123333. [CrossRef]
28. Chu, X.; Zhan, J.Y.; Li, Z.H.; Zhang, F.; Qi, W. Assessment on forest carbon sequestration in the Three-North Shelterbelt Program region, China. *J. Clean. Prod.* **2019**, *215*, 382–389. [CrossRef]
29. State Forestry and Grassland Administration. *China Forestry and Grassland Statistical Yearbook (2019)*; China Forestry Publishing House: Beijing, China, 2020; (In Chinese). ISBN 978-7-5219-0876-3.
30. Qin, H.Y. Research on the Relationship between Forest Ecology and Poverty in State-owned Forest Regions in Heilongjiang Province. Ph.D. Dissertation, Northeast Forestry University, Harbin, China, 2019. (In Chinese). [CrossRef]
31. Chen, J.; Gao, M.; Cheng, S.; Hou, W.; Song, M.; Liu, X.; Liu, Y.; Shan, Y. County-level CO_2 emissions and sequestration in China during 1997–2017. *Sci. Data* **2020**, *7*, 1–12. [CrossRef]

32. Chen, Y.; Feng, X.; Tian, H.; Wu, X.; Gao, Z.; Feng, Y.; Piao, S.; Lv, N.; Pan, N.; Fu, B. Accelerated increase in vegetation carbon sequestration in China after 2010: A turning point resulting from climate and human interaction. *Glob. Chang. Biol.* **2021**, *27*, 5848–5864. [CrossRef]
33. Zhang, X.; Brandt, M.; Tong, X.; Ciais, P.; Yue, Y.; Xiao, X.; Zhang, W.; Wang, K.; Fensholt, R. A large but transient carbon sink from urbanization and rural depopulation in China. *Nat. Sustain.* **2022**. [CrossRef]
34. Ding, Z.; Yao, S. Theory and valuation of cross-regional ecological compensation for cultivated land: A case study of Shanxi province, China. *Ecol. Indic.* **2022**, *136*, 108609. [CrossRef]
35. Lu, C.X.; Venevsky, S.; Shi, X.L.; Wang, L.Y.; Wright, J.S.; Wu, C. Econometrics of the environmental Kuznets curve: Testing advancement to carbon intensity-oriented sustainability for eight economic zones in China. *J. Clean. Prod.* **2021**, *283*, 124561. [CrossRef]
36. Li, C.; Fu, B.; Wang, S.; Stringer, L.C.; Wang, Y.; Li, Z.; Liu, Y.; Zhou, W. Drivers and impacts of changes in China's drylands. *Nat. Rev. Earth Environ.* **2021**, *2*, 858–873. [CrossRef]
37. Li, K.; Tong, Z.; Liu, X.; Zhang, J.; Tong, S. Quantitative assessment and driving force analysis of vegetation drought risk to climate change: Methodology and application in Northeast China. *Agric. For. Meteorol.* **2020**, *282–283*, 107865. [CrossRef]
38. Wu, D.; Xie, X.; Tong, J.; Meng, S.; Wang, Y. Sensitivity of Vegetation Growth to Precipitation in a Typical Afforestation Area in the Loess Plateau: Plant-Water Coupled Modelling. *Ecol. Modell.* **2020**, *430*, 109128. [CrossRef]
39. Jia, B.R.; Sun, H.R.; Shugart, H.H.; Xu, Z.Z.; Zhang, P.; Zhou, G.S. Growth variations of *Dahurian larch* plantations across northeast China: Understanding the effects of temperature and precipitation. *J. Environ. Manag.* **2021**, *292*, 112739. [CrossRef]
40. Luo, T.X.; Liu, X.S.; Zhang, L.; Li, X.; Pan, Y.D.; Wright, I.J. Summer solstice marks a seasonal shift in temperature sensitivity of stem growth and nitrogen-use efficiency in cold-limited forests. *Agric. For. Meteorol.* **2018**, *248*, 69–478. [CrossRef]
41. Vlam, M.; Baker, P.J.; Bunyavejchewin, S.; Zuidema, P.A. Temperature and rainfall strongly drive temporal growth variation in Asian tropical forest trees. *Oecologia* **2014**, *174*, 1449–1461. [CrossRef]
42. Peng, S.Z.; Ding, Y.X.; Liu, W.Z.; Li, Z. 1 km monthly temperature and precipitation dataset for China from 1901 to 2017. *Earth Syst. Sci. Data* **2019**, *11*, 1931–1946. [CrossRef]
43. Rogerson, P.A. Historical change in the large-scale population distribution of the United States. *Appl. Geogr.* **2021**, *136*, 102563. [CrossRef]
44. Wang, N.; Fu, X.; Wang, S. Spatial-temporal variation and coupling analysis of residential energy consumption and economic growth in China. *Appl. Energy* **2022**, *309*, 118504. [CrossRef]
45. Zuo, Z.; Guo, H.; Cheng, J.; Li, Y.L. How to achieve new progress in ecological civilization construction?—Based on cloud model and coupling coordination degree model. *Ecol. Indic.* **2021**, *127*, 107789. [CrossRef]
46. Hansen, B.E. Threshold effects in non-dynamic panels: Estimation, testing, and inference. *J. Econ.* **1999**, *93*, 345–368. [CrossRef]
47. Cai, Z.W.; Fang, Y.; Lin, M.; Su, J. Inferences for a partially varying coefficient model with endogenous regressors. *J. Bus. Econ. Stat.* **2019**, *37*, 158–170. [CrossRef]
48. Du, K.R.; Zhang, Y.H.; Zhou, Q.K. Fitting partially linear functional-coefficient panel-data models with Stata. *Stata J.* **2020**, *20*, 976–998. [CrossRef]
49. Geng, Y.D.; Sun, S.B.; Yeo-Chang, Y. Impact of forest logging ban on the welfare of local communities in northeast China. *Forests* **2021**, *12*, 3. [CrossRef]
50. Liu, S.L.; Xu, J.T. Livelihood mushroomed: Examining household level impacts of non-timber forest products (NTFPs) under new management regime in China's state forests. *For. Policy. Econ.* **2019**, *98*, 44–53. [CrossRef]

 forests

MDPI

Article

Using the Auction Price Method to Estimate Payment for Forest Ecosystem Services in Xin'an River Basin in China: A BDM Approach

Tan Li [1,*], BaoHang Hui [1], Le Zhu [1], Tianye Zhang [1], Tianyu Chen [1] and Chong Su [2]

[1] College of Economics & Management, Anhui Agricultural University, Hefei 230036, China; hbh@stu.ahau.edu.cn (B.H.); zhule0815@163.com (L.Z.); zty17805241625@163.com (T.Z.); chentianyu@ahau.edu.cn (T.C.)

[2] Institute of Agriculture Remote Sensing and Information Technology, College of Environmental and Resource Sciences, Zhejiang University, Hangzhou 310058, China; suchong@zju.edu.cn

* Correspondence: litan@ahau.edu.cn

Abstract: Accurately estimating the forest farmers' protection costs for forest ecosystem services has become a hot issue in ecological economics. In this research, we propose a novel method of using an auction price model to evaluate the forest ecosystem services. We establish a functional relationship between forest farmers and the forestland that belongs to them based on experimental data from Xin'an River Basin in China. The results indicate that the average willingness of farmers to accept payment for forest ecosystem service protection in the low, middle, and high levels of forest quality is 17,123.10, 23,493.75, and 31,064.40 yuan/ha/year, respectively. Moreover, farmers with different individual characteristics, household characteristics, planting characteristics, policy cognition, and ecological awareness are also willing to be paid differently. This research can provide a reference for forest ecosystem protection policies and assist the sustainable forestry development.

Keywords: auction price; payments for ecosystem services; Xin'an River Basin

Citation: Li, T.; Hui, B.; Zhu, L.; Zhang, T.; Chen, T.; Su, C. Using the Auction Price Method to Estimate Payment for Forest Ecosystem Services in Xin'an River Basin in China: A BDM Approach. *Forests* **2022**, *13*, 902. https://doi.org/10.3390/f13060902

Academic Editors: Elisabetta Salvatori and Giacomo Pallante

Received: 2 May 2022
Accepted: 8 June 2022
Published: 9 June 2022

1. Introduction

Providing forest ecosystem service performance rewards is an alternative to command-and-control or indirect conservation incentives [1–3]. By compensating forest owners for conservation costs, the private and public benefits of conservation can be reconciled [4–9]. The use of payment requires accurate estimations of private costs, in particular the willingness of forest owners to accept protection contracts [10–14]. This insufficient compensation will not result in behavioral changes due to the high violation rates for forest owners [15]. If these owners are not compensated accurately, the approach would not maximize conservation benefits [16–18].

The basic concept of forest ecosystem services is defined as public goods that are easily overused and underestimated, which will lead to the loss of service value, exploitation, and fragmentation, and seriously reduce service functions [19–21]. Therefore, forest ecological compensation is of great importance for the rational use of forest resources and the effective protection of forest ecology [22–24]. According to the policy, stakeholders' interests are adjusted by means of economics and are compensated for the costs (or losses caused by the destruction of resources and the environment) of their protection, to achieve the purpose of preserving the ecosystem [25–27]. Payment for Ecosystem Services (PES) is a project tool that is widely used internationally [28–30]. PES projects compensate for the reconstruction of ecosystem services caused by improving the damaged environmental conditions [31–34].

A literature review of ecological economics shows that conservation benefits can be obtained by incorporating cost measures into conservation planning processes [23]. Nevertheless, conservation planners do not know what the opportunity cost of conservation is for forest owners. Consequently, there are a variety of methods for estimating these costs [35,36].

In the past few years, scholars found that the current payment option in PES projects cannot accommodate information asymmetry, which can result in the loss of some protectors. A lack of adequate compensation leads to an uneven distribution of compensation funds and social injustice [37]. Thus, some studies use auctions to deal with information asymmetry [38]. In contrast, the Contingent Valuation Method (CVM) [39,40] and Choice Experiments (CE) [41,42] are widely used and highly recognized. A CVM method was first proposed and used by Davis in 1963 to evaluate the recreational value of forestland in Maine [43]. In recent years, the CVM method has been widely applied to measure both the use value and non-use value of environmental goods. However, there are some limitations, such as the fact that only one environmental attribute can be measured at one time [44] and that inaccurate results can easily occur as a result of inherent bias [43]. A CE method is based on Lancaster's characteristics theory of value [45] and random utility theory [46,47]. Originally developed by Louvier [48,49] and others [50], it allows participants to establish a selection set by setting different attribute combinations so that each attribute can be weighed and the most preferred combination scheme can be chosen. Nevertheless, both the CE and the CVM are essentially based on hypothetical markets. In the hypothetical market, it is common for consumers to overestimate or underestimate their actual willingness to pay (or accept) for goods and services. Since the ecological compensation program in China will establish a market mechanism for adjusting interests and allocating resources, experimental economics should be used to simulate the market environment and achieve compensation standards that are more adaptable with the market prices of forest ecosystem services.

The auction price method is widely used as a representative method to evaluate commodities in non-market transactions. The effectiveness of experimental auction methods depends mainly on the choice of auction mechanism. Vickrey auction mechanisms [50], BDM (Becker-DeGroot-Marschak) mechanisms [51], and random n-price auction mechanisms [52] are currently widely used. Vickrey auctions require all participants to bid simultaneously. The highest bidder wins and only has to pay the second highest, which generally satisfies the principle of incentive compatibility, although there may be an issue with low bids in the implementation and cannot be implemented in individual experiments. BDM does not require group participation and it is suitable for individual experiments, which provides the assurance of randomness in sampling, as well as the avoidance of the information association defect caused by group auctions. Thus, the "insincere" nature of bidding can be effectively eliminated by Random n-price auctions, which combine the advantages of Vickrey auctions and BDM mechanisms and comply with the principle of incentive compatibility, overcome competitive bias, and obtain the most accurate data [53]—but they have the disadvantage of being difficult to explain to participants and difficult to organize. In the experiment, the auction mechanism is introduced and the participants' real preferences can be gathered; the rigorous experimental procedures and the repeated experimental process make the experimental environment more like the real market: the participants can be familiar with the auctioned items and the experimental process ensures data authenticity and prevents the problem of non-response bias. Recent studies have concentrated on the design and cost-effectiveness of ecological service auctions. Lewis et al. have tried to build an auction mechanism in the context of climate change [54]. Sharma et al. have examined forest carbon sequestration service payments under discriminatory price auctions [55]. Lundberg et al. have explored fixed payments and the procurement cost-effectiveness of payments for ecosystem services under two mechanisms of auction [28].

Therefore, we investigate the minimum price they require for the contract activities for the provision of forest ecosystem services using an experimental auction method based on the BDM mechanism. Moreover, we establish the multivariant probit model (MVP), which aims to obtain the farmers' characteristics, household characteristics, policy cognition, and ecological cognition to explore differences in the farmers' willingness to accept a contract (WTA) for forest ecosystem services at different quality levels. Additionally, we compare their influencing factors. Finally, we calculate a reasonable range of ecological compensation standards. This research makes the following contributions to the previous literature: (i) exploring the underlying theory of forest ecosystem service measurement and its connection with farmers. In this study, we discuss the auction price method and the advantages of the BDM approach compared to the conventional methods; (ii) regarding real marketing among farmers, it is meaningful to explain the differences in economic value that are attributable to the different farmers' groups; and (iii) it is expected that the analytical results will have implications for forest ecosystem management regarding ecological transfer payments and its budget allocation.

In the following section, the paper is organized as follows. Section 2 proposes an empirical model framework and outlines the experimental design and data collection process. Section 3 addresses the model results, and is followed by Section 4, which presents the study summary and conclusions.

2. Methodology

2.1. Sampling and Research Area

We designed a questionnaire that involved two aspects: one related to the individual characteristics, household and planting characteristics, and the social capital of farmers. The other regarded simulating the form of forestland lease auction contract. The experiment was conducted in two groups for household research. Firstly, the head of the household (or a family member who knew about the production of forestland) was asked to fill out a basic questionnaire. Then, the researcher introduced and briefly demonstrated the auction processes and rules. In the end, the experiment was conducted by simulating forestland lease auction contracts. A total of 300 farmers were invited for this experiment, with a valid sample size of 272. The efficiency of the questionnaire reached 90.67%.

As shown in Figure 1, the study sample covered ten villages in Anhui province in China, and thus represented a typical range of conditions in the ecological management of the Xin'an River and its core areas for implementing an ecological compensation policy. The whole survey process lasted about six months—from June to December in 2020. The sample in our study was randomly selected from Xiuning County and She County of Huangshan City. Xiuning County is the fountainhead of Xin'an River. She County is located at the junction of Anhui and Zhejiang province, where the water quality data of the station in Jiekou Town is a key judgement criterion for compensation implementation. For each county, we selected 2–3 townships and 2–3 villages, which were randomly selected from the selected townships. Furthermore, we investigated 30 farmers in each village.

2.2. Auction Method

BDM auctions can better reveal the willingness preferences of farmers for its uses with real monetary incentives by setting up a real market environment. In order to achieve more accurate results, the whole experimental process must be rigorously designed. The experiment includes several components, such as determining the auction product, selecting the auction mechanism, and determining the experimental environment and execution steps. Each step must be carefully considered to guarantee the validity of the experimental auction results. The specific design framework is shown in Figure 2.

2.2.1. Auction Products

Forest ecosystem services, as intangible ecological products, are difficult to understand for farmers unless they are directly used as the auction products. Therefore, the selection of reasonable alternatives has a key impact on the experimental auction. Xiuning County and She County are located in the mountainous areas of southern Anhui, with undulating terrain and limited arable land. Farmers take forestry activities (tea and fruit trees) as their main income resources. At the end of 2019, the tea garden area of the two counties was 10,751 hectares and 17,142 hectares, respectively, accounting for 20.69% and 32.99% of the total tea garden area of Huangshan City, respectively. This study is based on the natural and economic situation of this area, and then combined with the existing research. In addition, the basic idea of the equivalence factor method is applied to the experiment. It should be noted that the auction products need not only reflect the value of forest ecosystem services, but also the differences in farmers' characteristics. Hence, the one-year use right of farmers' operating on forestland is selected as the auction product. The bid of rent is used as the compensation price for the value of forest ecosystem services per unit area when farmers participate in ecological compensation projects.

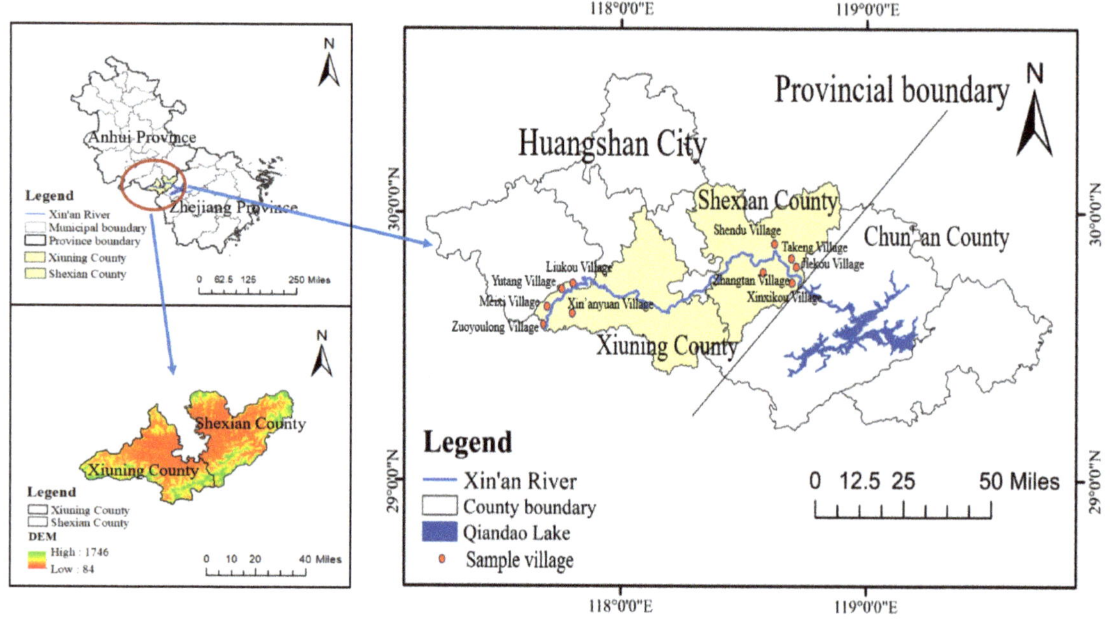

Figure 1. Study area in China.

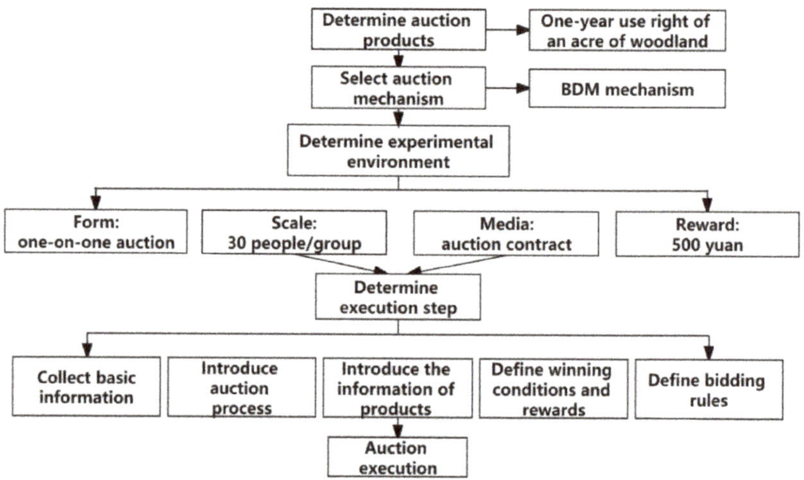

Figure 2. Auction experimental design framework.

In addition, in order to simulate the actual market environment, we classified forestland into high quality, middle quality, and low quality, including the expression of the farmer as well as the slope of the land, fertility, irrigation, and transport conditions (Table 1). Bidders needed to report their bids for the lease amount of forestland of different qualities during the experimental auction, respectively. We could explore farmers' willingness to accept (WTP) for different levels of forest ecosystem services based on their bids.

Table 1. Classification of forest quality levels.

Woodland Levels	Woodland Slope	Soil Fertility	Irrigation Conditions	Transport Condition	Variety
Low quality forest land	Steeper (>25°)	Thin soil layer, low fertility	No water around.	Country road (for motorcycles and tricycles)	Poor tea varieties
Middle quality forest land	Stee ($15° \leq$ slope $\geq 25°$)	Low fertility	The water source is far away but there are water diversion facilities.	Close to township roads (for small cars)	Tea varieties in general
High quality forest land	Gentle (<15°)	Thick soil layer, high fertility	Close to the water source (within 100 m)	Close to the highway (for large trucks)	Good tea varieties

2.2.2. Auction Selection Mechanism

For it is difficult to organize large-scale centralized auctions in the scattered households in mountainous areas, this study conducts the experimental auction by simulating a forestland lease contract in a household. Without reference to others, the farmer made an independent valuation of the auction products. This could be seen as organizing a sealed auction [32]. In addition, due to the establishment of three levels of auction products, the auction was a multi-item auction. A BDM mechanism was chosen as the experimental auction mechanism to prevent the "insincerity" of farmers' bids and take into account the "randomness" of sampling. We set up a normally-distributed function based on the bid interval of participants. In each round of different levels of auctions, if the farmers' bids are lower than the randomly selected price, the forest lease contract will be effective, otherwise it will fail. Moreover, the BDM mechanism does not require group participation. It is suitable for individual experiments with a stratified random sampling method. Meanwhile, under the BDM mechanism, participants have a chance to win regardless of their valuation.

2.2.3. Auction Procedures

Considering that the natural environment and socio-economic conditions vary greatly from different regions, local farmers' production operations will also vary widely. Therefore, this study took administrative villages as a basic unit. We selected 30 farmers in each village randomly to conduct a three-round sealed auction with one-on-one household offers. Briefly, the BDM auction was implemented as follows: firstly, in each round of the auction, farmers needed to bid on the corresponding level of forestland in this round. Then, a normal-distribution random function generated the transaction price for this round. If the farmer's bid for this round was lower than or equal to the randomly generated transaction price, the transaction would be concluded. After three rounds of auction, the computer randomly selected one round from the three rounds of auctions as the final settlement round. If the farmers could reach a deal in the settlement round, they would be included in the winning group. Then, the farmers in the winning group were ranked in order of their bids. The farmer with the highest bid won and received a cash prize of 500 yuan. The related experimental procedure is shown in Figure 3.

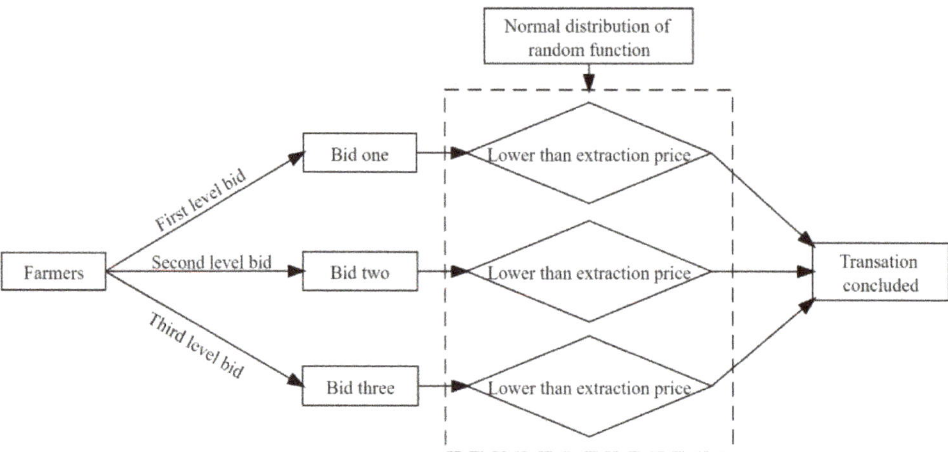

Figure 3. BDM mechanism program.

2.3. *Statistical Modeling*

2.3.1. Multivariant Probit (MVP) Model

According to Lancaster's utility theory [45], let U_{ij} be the utility of the attribute obtained by the ith farmer for selecting the jth level of forest ecosystem service, including two components. The first is the deterministic component V_{ij}, the second is the stochastic term ε_{ij}, as below:

$$U_{ij} = V_{ij} + \varepsilon_{ij} \tag{1}$$

Moreover, if the market value of forest ecosystem services at the jth level is P_j, the remaining willingness of the ith farmer to be paid for forest ecosystem services at the jth level is AP_{ij}, which can be expressed as:

$$AP_{ij} = V_{ij} - P_j + \varepsilon_{ij} \tag{2}$$

Based on the incentive compatibility properties of the revealed preference approach and the BDM mechanism, there exists $BID_{ij} = WTP_{ij} = V_{ij}$. BID_{ij}, which is the bid that the ith farmer wants for the jth level of forest ecosystem services. In addition, since the market price P_j of forest ecosystem services does not have an exact value, the average value $\overline{WTA}_{ij}$ of farmers' bids for different levels of forest ecosystem services is used as a proxy for the market price P_j, which is shown as below:

$$AP_{ij} = WTA_{ij} - \overline{WTA}_{ij} + \varepsilon_{ij} \tag{3}$$

$\overline{WTA}_{ij}$ is the arithmetic mean of all farmers for the ecological services WTA_{ij} at the jth level and ε_{ij} is a random parameter. If $AP_{ij} \geq 0$, then the ith farmer desires a higher compensation for the ecological service at the jth level and vice versa. Accordingly, a binary discrete choice model is constructed:

$$Y_{ij} = \begin{cases} 1\, AP_{ij} \geq 0 \\ 0\, AP_{ij} < 0 \end{cases} \tag{4}$$

$Y_{ij} = 1$ denotes a high willingness of the ith farmer to be paid for the jth level of ecosystem services, otherwise $Y_{ij} = 0$. It can be further expressed as:

$$AP_i = X_i\beta + \varepsilon_i \tag{5}$$

In addition,

$$X_i = \begin{bmatrix} X_{i11} \cdots X_{i1m} & & & \\ & X_{i21} \cdots X_{i2m} & & \\ & & \cdots\cdots & \\ & & & X_{i11} \cdots X_{i1m} \end{bmatrix} \tag{6}$$

X_{ijm} denotes that the mth characteristic variable of the ith farmer in the jth bid. β is the parameter vector to be estimated and ε_i is the residual term.

Therefore, the probability that the farmer hopes to obtain a higher WTA for the jth level of ecological services is calculated as:

$$\begin{aligned} Prob(Y_i = 1) &= Prob(AP_i \geq 0) = F(\varepsilon_i - X_i\beta) \\ &= 1 - F(-X_i\beta) \end{aligned} \tag{7}$$

If ε_i obeys, the normal distribution is obeyed and the assumptions of the MVP model are satisfied, so the function is:

$$Prob(Y_i = 1) = 1 - \Phi(-X_i\beta) = \Phi(X_i\beta) \tag{8}$$

There are three hierarchical subjects in this study, so $j = 3$. Since the MVP model assumes that the residual terms obey a joint normal distribution, therefore $\varepsilon_i \sim N(0, \Sigma)$, then $AP_i \sim N(X_i\beta, \Sigma)$, where $\Sigma = \begin{bmatrix} 1 & \sigma_{12}\sigma_{13} \\ \sigma_{12} & 1 \ \sigma_{23} \\ \sigma_{13}\sigma_{23} & 1 \end{bmatrix}$.

If the three bids are not correlated, then $\sigma_{12} = \sigma_{23} = \sigma_{13} = 0$. To detect the correlation of the three bids, this study assumes that σ_{12},σ_{23}, and σ_{13} are not zero [35]. Then, based on the benchmark model proposed by Chib et al. [36], the MVP model can be obtained as:

$$\begin{aligned} Prob(Y_i|\beta, \Sigma) &= Prob(Y_i, AP_i|X_i\,\beta, \Sigma) \\ &= \int_{B_{i3}}\int_{B_{i2}}\int_{B_{i1}} \varphi(Y_i, AP_i|X_i\,\beta, \Sigma)dAP_i \end{aligned} \tag{9}$$

$\varphi(Y_i, AP_i) = \frac{1}{(2\pi)^{3/2}|\Sigma|^{1/2}} e^{\frac{1}{2}}(AP_i - X_i\beta)'\Sigma^{-1}(AP_i - X_i\beta)$ is the joint probability density function and B_{ij} is the integration interval.

$$B_{ij} = \begin{cases} (0, +\infty)\, Y_{ij} = 1 \\ (-\infty, 0)\, Y_{ij} = 0 \end{cases} \tag{10}$$

The likelihood function of the model can be obtained as:

$$\mathrm{L}(\theta) = \prod_{i=1}^{272} \varphi(Y_i, \Delta AP_i|\beta, \Sigma) \tag{11}$$

The log-likelihood function is:

$$\ln(\mathrm{L}(\theta)) = \ln(\prod_{i=1}^{272} \varphi(Y_i, AP_i|\beta, \Sigma) = \sum_{i=1}^{272} \ln\{\varphi(Y_i, AP_i|\theta)\} \tag{12}$$

while $\theta = (\beta, \sum)$ is the parameter space.

2.3.2. Variable Definition

Referring to the previous literature [15,20], this study used "whether the bid for ecological services at three different levels was higher than the average of all bids at each time" as the dependent variable. In order to investigate the influencing factors of farmers' WTA for ecological services at different levels, independent variables from multiple aspects were set up as following: (1) farmers' characteristics variables, including gender, age, education, health status, and occupations; (2) farmers' household characteristics and planting characteristics variables, including per capita income, household size, forestland area, average cost per hectare, and average output per hectare, etc; and (3) farmers' policy perceptions and ecological perceptions, including policy cognition, policy satisfaction, policy support, and ecological awareness. The details are shown in Table 2.

Table 2. Variable definitions and assignments.

Variable		Definition	Mean	Std. Dev.
Dependent variable				
	Higher than average price for low quality forest land	1 = the bid higher than the average price; 0 = otherwise	0.338	0.474
	Higher than average price for medium quality forest land	1 = the bid higher than the average price; 0 = otherwise	0.342	0.475
	Higher than average price for high quality forest land	1 = the bid higher than the average price; 0 = otherwise	0.272	0.446
Independent variable				
Individual characteristics	Gender	1 = male, 0 = female	0.636	0.482
	Age	1 = Over 40 years old, 0 = otherwise	0.952	0.214
	Education	1 = High School and above, 0 = otherwise	0.114	0.318
	Health	1 = healthy; 0 = chronic diseases, major diseases and disabilities	0.676	0.469
	Occupation	1 = farmer, 0 = otherwise	0.809	0.394
Family characteristics	Income	Continuous variable/1000 yuan	10.889	9.649
	Household size	Total household size, continuous type variable/person	4.239	2.043
Planting characteristics	Forest area	Total area of operating forest land, continuous type variable/ha	0.315	0.259
	Average cost per hectare	Average cost per hectare of operating forest land, continuous type variable/1000 yuan	6.075	7.380
	Average output per hectare	Average output per hectare of operating forest land, continuous type variable/1000 yuan	31.800	21.015
Policy Awareness	Policy cognition	1 = Familiarity, 0 = otherwise	0.739	0.440
	Policy Satisfaction	1 = Satisfaction, 0 = otherwise	0.794	0.405
	Policy Support	1 = Support, 0 = otherwise	0.960	0.197
Ecological Awareness	Ecological Awareness	1 = Necessity, 0 = otherwise	0.886	0.318

3. Results

3.1. Descriptive Statistics

Table 3 summarizes the characteristics of the farmers. Respondents were familiar with the PES policy and satisfied with the PES Programs in the Xin'an River. Most respondents both support the policy and are aware of the ecological environment. As can be seen in Table 1, the demographic characteristics of the samples are very similar to the local statistics data. Our sample is representative of the education composition of the true population, but has a higher mean age compared to the general population. The results demonstrate that the proportion of males is higher than that of females, accounting for 63.60% and 36.40%, respectively. The age of the participants is mainly above 40-years old. It reflects that the local farmers are mainly middle-aged and elderly.

Table 3. Statistics of basic characteristics of peasant households.

Statistical Characteristics	Classification Indicators	Number	Proportion %
Gender	male	173	63.60
	female	99	36.40
Age	<40	13	4.78
	≥40	259	95.22
Education	Junior high school or below	241	88.60
	High school or junior college degree or above	31	11.40
Health Conditions	Healthy	201	67.65
	Chronic diseases, major diseases and disabilities	71	32.35
Occupation	Farmer	220	80.88
	Others	52	19.12
Policy Cognition	Familiarity	201	73.90
	Unfamiliarity	71	26.10
Policy Satisfaction	Satisfaction	216	79.41
	Dissatisfaction	56	20.59
Policy Support	Support	261	95.96
	Nonsupport	11	4.04
Ecological Awareness	Necessity	241	88.60
	Unnecessary	31	11.40

In Table 4, we can see that the annual average cost range of farmers operating forestland is [0, 43,500] yuan/ha, the annual average output per hectare range is [0, 112,500] yuan/ha, and the annual average net income per hectare range is [−2250, 97,500] yuan/ha. In addition, the annual average cost is 6075 yuan/ha, the annual average output is 31,800 yuan/ha, and the annual average net income is 25,725 yuan/ha.

Table 4. Basic situation of farmers' forestland.

	Average Cost (Yuan/Ha/Year)	Average Output (Yuan/Ha/Year)	Average Net Income (Yuan/Ha/Year)
Max	43,500	112,500	97,500
Min	0	0	−2250
Mean	6075	31,800	25,725

3.2. Bids Prices in Experiments

The basic information of farmers' bids for forestland of different quality levels is shown in Table 5. It can be seen that the range for different quality levels of forestland are low-quality forestland ([0, 112,500] yuan/ha/year), middle-quality forestland ([1500, 150,000] yuan/ha/year), and high-quality forestland ([7500, 225,000] yuan/ha/year), respectively. Furthermore, the mean offers submitted at different levels are 17,123.1, 23,493.75, and 31,064.40 yuan/ha/year, respectively. The spillover of farmers' bids compared with the average output are −28.48%, −1.65%, and 28.09%, respectively. Compared to the average net income, the spillover prices are −14.16%, 18.54%, and 54.74%, respectively.

The bids for middle-quality forestland are lower than the actual output of forestland and higher than the net income of forestland, which matches the rules of marketing. However, the bids for low-quality forestland are much lower than the actual output and net income of forestland. These farmers are willing to contract to others for free and only require that the use of the forestland not fall into waste. This may be because the profit of low-quality forestland is low. It is not convenient for farming operations such as planting and irrigation, and part of the land is already in a semi-deserted state. As expected, farmers generally charge higher prices for high-quality forestland, which is much higher than the actual output and net income of the forestland. Due to the harsh natural conditions in mountainous areas, high-quality forestland is a scarce resource in the local area. The tea revenues are a major source of income for farmers. It also indicates that farmers have a strong relationship with the forestland and are unwilling to lease out high-quality forestland. Compared to the forestland market in our survey, the actual transaction price is 30,000–90,000 yuan/ha/year.

Table 5. Basic situation of auction quotation.

	Low-Quality	Middle-Quality	High-Quality
Highest bid	112,500.00	150,000.00	225,000.00
Average bid	17,123.10	23,493.75	31,064.40
Lowest bid	0.00	1500.00	7500.00
Bid and average premium for total output	−28.48%	−1.65%	28.09%
Bid and average premium for net income	−14.16%	18.54%	54.74%

3.3. Model Analysis

We analyze the survey data using STATA 15.0, and the results are shown in Table 6. $\sigma12$, $\sigma13$, and $\sigma23$ are significant and not a zero value. This indicates that the three bids of farmers are highly correlated, which is suitable for MVP model analysis.

3.3.1. Influence of Farmers' Characteristics

The results of the model demonstrated that gender and education level have no influence on WTA. Given that they are probably influenced by mountainous terrain and economic conditions, the education level of local farmers is generally low. Moreover, due to the implementation requirements and difficult operation of the experimental auction method, the sample size we can obtain is small. Given this, it has a certain impact on the regression results. The results also show that age, health, and job type have significant effects on the three levels of WTA. From the aspect of age, there is no difference on WTA between farmers aged over 40 and those aged below 40 in the first level of ecological service. While the WTA of farmers who are above 40 years old is significantly higher in the second and third levels, the reason may be that the middle-aged and elderly people's own labor ability is limited; they lack employment opportunities and depend on the forest land income. Hence, they hope to receive higher compensation. Our results also show that, for second-level ecological services, farmers in good health hope to achieve more compensation. It may be that the healthier the farmers are, the stronger their labor capacity and the higher yield from production work. In terms of occupations, the WTA of farmers working in forestry was significantly higher than that of farmers working in non-forestry work in the first and second levels. We conclude that farmers working in forestry are highly dependent on land and forestry income. Hence, they want more compensation in order to make up for the losses caused by the implementation of the ecological compensation policy.

3.3.2. Influence of Household and Planting Characteristics

Among the farm household characteristics, household per capita income has no significant difference on WTA for the three levels of ecosystem services. We find that farmers with larger household sizes have significantly higher WTA, except in the second level. It is possible that the larger the household size, the higher the cost of living. Farmers wish to obtain higher compensation for their families to gain benefits.

There is no significant difference in WTA between farmers with different forest area, and there is significant difference between farmers with different costs and outputs. Firstly, there is no significant difference in the WTA of the first level of farmers with different costs, and the WTA of the second and third levels of farmers with higher costs is significantly lower than that of farmers with low costs. The reason is that, as the average cost is high, the general planting scale is small. Therefore, the WTA of farmers is lower. Secondly, the WTA of farmers with high output is significantly higher than that of farmers with low output. The reason is that, when the level of land output is high, farmers have a higher income and they perceive a higher ecosystem service value.

Table 6. MVP model results.

	Variable	Coefficient	Std. Dev.	Z-Value	*p*-Value
Low quality	Gender	−0.071	0.188	−0.380	0.705
	Age	0.185	0.459	0.400	0.687
	Education	−0.063	0.290	−0.220	0.829
	Health Conditions	0.326 *	0.195	1.670	0.095
	Occupation	0.730 ***	0.250	2.920	0.003
	Income	−0.002	0.009	−0.220	0.823
	Household size	0.079 *	0.044	1.810	0.070
	Economic forest area	−0.021	0.027	−0.760	0.449
	Average land cost per hectare	−0.007	0.216	−0.030	0.975
	Average income per hectare of land	0.407 ***	0.084	4.840	0.000
	Policy cognition	−0.548 ***	0.198	−2.770	0.006
	Ecological Awareness	0.592 **	0.292	2.030	0.042
	Policy Satisfaction	0.543 **	0.234	2.320	0.020
	Policy Support	−0.203	0.509	−0.400	0.690
	_cons	−2.894 ***	0.852	−3.400	0.001
Middle quality	Gender	−0.061	0.168	−0.360	0.716
	Age	1.141 ***	0.411	2.770	0.006
	Education	0.059	0.263	0.220	0.822
	Health Conditions	0.269	0.174	1.550	0.121
	Occupation	0.347 *	0.205	1.700	0.090
	Income	0.005	0.009	0.580	0.563
	Household size	0.050	0.040	1.230	0.217
	Economic forest area	−0.010	0.022	−0.430	0.668
	Average land cost per hectare	−0.415 **	0.209	−1.980	0.047
	Average income per hectare of land	0.329 ***	0.071	4.620	0.000
	Policy cognition	−0.297 *	0.171	−1.740	0.082
	Ecological Awareness	0.253	0.254	1.000	0.319
	Policy Satisfaction	0.645 ***	0.218	2.960	0.003
	Policy Support	−0.551	0.436	−1.260	0.207
	_cons	−2.756 ***	0.738	−3.730	0.000
High quality	Gender	−0.180	0.172	−1.040	0.297
	Age	1.200 ***	0.436	2.750	0.006
	Education	−0.203	0.253	−0.800	0.422
	Health Conditions	0.455 **	0.181	2.520	0.012
	Occupation	0.234	0.203	1.150	0.249
	Income	0.006	0.008	0.750	0.454
	Household size	0.068 *	0.038	1.760	0.078
	Economic forest area	−0.011	0.022	−0.500	0.615
	Average land cost per hectare	−0.355 *	0.202	−1.760	0.079
	Average income per hectare of land	0.294 ***	0.068	4.350	0.000
	Policy cognition	−0.301 *	0.167	−1.800	0.072
	Ecological Awareness	0.509 *	0.270	1.890	0.059
	Policy Satisfaction	0.636 ***	0.222	2.870	0.004
	Policy Support	−0.866 **	0.424	−2.040	0.041
	_cons	−2.873 ***	0.748	−3.840	0.000
	σ_{12}	0.933 ***	0.021	43.630	0.000
	σ_{13}	0.856 ***	0.031	27.540	0.000
	σ_{23}	0.972 ***	0.011	86.380	0.000
	Likelihood ratio test of rho12= rho13 = rho23 = 0: chi2(3) = 276.606 Prob > chi2 = 0.0000 R2 = 16.05				

Note: *, **, *** represent significant at the 10%, 5%, and 1% levels, respectively.

3.3.3. Influence of Farmers' Policy and Ecological Perceptions

The results show that different policy cognition, policy satisfaction, and policy support have a significant influence on WTA. In terms of policy cognition, farmers who are familiar with the ecological compensation policy have a significantly lower WTA. Typically, if farmers are familiar with this policy, they will be confident in policy implementation and have a lower estimate of risk loss. With respect to policy satisfaction, farmers who are satisfied with the policy implementation have a significantly higher WTA. The reason may be that their estimation of the value of ecological services is correspondingly increased along with their satisfaction. Regarding policy support, farmers who support policies have a lower WTA, suggesting that they are more likely to recognize the long-term benefits of policies and are willing to abandon certain economic benefits to improve the forest ecosystem services.

Farmers who think ecological protection is necessary exhibit significantly higher than farmers who do not in both the low and high levels. It may be that the stronger the farmers' awareness of ecological conservation, the higher the estimation of the value of ecological services; they prefer to put ecological conservation into practice. It also indicates that the farmers who rent forestland with low and high levels in this area are more sensitive than those who rent middle-level forestland.

4. Discussions, Conclusions and Policy Implications

4.1. Discussions

Basically, ecological compensation standards are established based on the WTA. The upper limit can be used as the compensation standard agreed upon by the respondent. Our study showed how an auction price model can be used to estimate the compensation standard in a PES program. Even in the absence of well-functioning markets, the auction price approach overcomes the weaknesses of existing valuation methodologies. Analyzing the potential impacts of different PES targeting programs provides the opportunity to achieve ecological and socio-economic goals.

The value of the research extends beyond the limitations of local considerations. By conducting similar auctions at different locations and for multiple services around the globe, scientists, practitioners, and policy-makers can gain a better understanding of what budget is necessary to pay for ecosystem services on a global scale. Additionally, our experimental auction price design will reduce the opportunity costs of environmental conservation. Field auctions, for instance, may be used to illustrate whether educating or influencing factors can lower the opportunity costs farmers face when using ecosystem services. As a result, we recommend continued experimentation with auction price models as a method of revealing the preference for PES program design, which is necessary and will improve the success of ecosystem service conservation.

According to the research, the evaluation of the effectiveness of PES policies should be based on the preferences of stakeholders, and only policies that meet the needs of most stakeholders are effective. It is important to note that although the experimental auction method is consistent with the actual WTA in theory, it requires further confirmation in the trading market.

4.2. Conclusions

This study uses a Becker-DeGroote-Marshack (BDM) auction method to estimate the willingness to accept (WTA) for different levels of ecosystem services. In the background of the implementation of Xin'an River ecological compensation policy, this study explores the accuracy of the forest ecological compensation standard in Huangshan City. Subsequently, we investigate the differences in farmers' preferences for different levels of ecosystem services and the factors affecting them with a sample of 272 farmers.

In Table 7, the auction experiment indicates that farmers have a lower average bid for low-quality forestland (17,123.1 yuan/ha/year) and a higher average bid for middle-quality forestland (23,493.75 yuan/ha/year) and high-quality forestland (31,064.4 yuan/ha/year). The BDM mechanism implements a one-to-one multi-round sealed auction that requires realistic monetary incentives and simulates a real market environment. Such an auction mechanism allows farmers' bids to truly reflect their preferences for forest ecosystem services. Given this, their bids can be an important reference for ecological compensation standards.

Table 7. Summary of calculation results (Unit: Yuan/ha/year).

	Farmers' Bids in Experimental Auction			
	Low Quality	**Middle Quality**	**High Quality**	**Average**
price	17,123.10	23,493.75	31,064.40	23,893.80

The MVP model is constructed to analyze the differences in farmers' bids and the factors influencing their preferences. The results show that there are significant differences on WTA among farmers with different individual characteristics, household characteristics, planting characteristics, policy cognition, and ecological awareness. Age, health status, and occupations have significant positive effects on farmers' WTA. In terms of the family characteristics, household size is found to have a positive effect on farmers' WTA. With planting characteristics, the average cost and average output have significant negative and positive effects on WTA, respectively. Policy cognition and policy satisfaction have significant positive effects on farmers' WTA. Policy support has a significant negative effect on WTA. Ecological awareness has a significant positive effect on WTA.

Lastly, this study takes the lowest of the average bid of farmers at each level as the lowest limit of the compensation standard and the highest as the upper limit of the compensation standard. As a result, we deduce that the reasonable range of the current forest ecological compensation standard in Huangshan City is [17,123.1, 31,064.4] (yuan/ha/year).

4.3. Policy Implications

This research provides three main implications to policymakers.

Farmers generally reflect that the current low standard of ecological compensation is difficult to make up for the loss of economic benefits by the limited development of livelihoods. This indicates that the existing compensation standard should be raised and government departments should implement diversified compensation methods. Our findings indicate that farmers generally have a higher acceptance of non-direct monetary compensation. They also have a higher willingness to participate in technical training, industrial support, and other projects. Thus, local governments should pay attention to farmers' anticipation and adopt diversified compensation methods, such as employment technology training and improving village infrastructure conditions.

The results show that farmers' WTA and their willingness to participate in policy are strongly related to their policy cognition and ecological consciousness. It is advisable that education should be strengthened to improve farmers' policy awareness and ecological consciousness. Due to the constraints of natural geography and socio-economic conditions in mountainous areas, farmers are less educated to obtain more information. Given this, their average policy awareness and ecological consciousness are limited. We can fundamentally improve the policy cognition and ecological awareness of farmers and promote ecological compensation policy implementation by increasing education and popularizing policy information and ecological knowledge.

Lastly, it is imperative to create a market-based compensation mechanism and introduce social capital. As an important method for allocating resources and determining prices, the auction mechanism has been widely implemented in PES projects worldwide. In China, the auction mechanism has been used in many fields such as emission rights, water rights, and land-use rights. Meanwhile, this study has tried the application of an auction mechanism and the results are more valid and accurate than other revealed preference methods. Hence, it is highly feasible to apply the auction mechanism in ecological compensation in developing countries.

Author Contributions: Conceptualization, T.L. and T.C.; methodology, T.C.; software, B.H. and L.Z.; validation, L.Z., B.H. and T.C.; formal analysis, B.H. and L.Z.; investigation, B.H. and T.Z.; writing—original draft preparation, B.H. and L.Z.; writing, review and editing, T.L. and C.S.; project administration, T.L.; funding acquisition, T.L. All authors have read and agreed to the published version of the manuscript.

Funding: This research was funded by the National Natural Science Foundation of China, grant number 71873003, 71503004.

Data Availability Statement: No new data were created or analyzed in this study. Data sharing is not applicable to this article.

Conflicts of Interest: The authors declare no conflict of interest.

References

1. Jack, B.K.; Leimona, B.; Ferraro, P.J. A Revealed preference approach to estimating supply curves for ecosystem services: Use of auctions to set payments for soil erosion control in Indonesia. *Conserv. Biol.* **2009**, *23*, 359–367. [CrossRef] [PubMed]
2. Engel, S.; Pagiola, S.; Wunder, S. Designing payments for environmental services in theory and practice: An overview of the issues. *Ecol. Econ.* **2008**, *65*, 663–674. [CrossRef]
3. Muradian, R.; Corbera, E.; Pascual, U.; Kosoy, N.; May, P. Reconciling theory and practice: An alternative conceptual framework for understanding payments for environmental services. *Ecol. Econ.* **2010**, *69*, 1202–1208. [CrossRef]
4. Pattanayak, S.K.; Wunder, S.; Ferraro, P.J. Show me the money: Do payments supply environmental services in developing countries? *Rev. Environ. Econ. Policy* **2010**, *4*, 254–274. [CrossRef]
5. Bo, S.; Yue, Z.; Lixiao, Z.; Fan, Z. A top-down framework for cross-regional payments for ecosystem services. *J. Clean. Prod.* **2018**, *182*, 238–245.
6. Isabel, V. Payments for ecosystem services in the context of adaptation to climate change. *Ecol. Soc.* **2012**, *17*, 11.
7. Jindal, R.; Kerr, J.M.; Ferraro, P.J.; Swallow, B.M. Social dimensions of procurement auctions for environmental service contracts: Evaluating tradeoffs between cost-effectiveness and participation by the poor in rural Tanzania. *Land Use Policy* **2013**, *31*, 71–80. [CrossRef]
8. Ingram, J.C.; Wilkie, D.; Clements, T.; Mcnab, R.B.; Foley, C.A.H. Evidence of payments for ecosystem services as a mechanism for supporting biodiversity conservation and rural livelihoods. *Ecosyst. Serv.* **2014**, *7*, 10–21. [CrossRef]
9. Brockerhoff, E.G.; Barbaro, L.; Castagneyrol, B.; Forrester, D.I.; Gardiner, B.; Gonza'lez-Olabarria, J.R.; Lyver, P.O.B.; Meurisse, N.; Oxbrough, A.; Taki, H.; et al. Forest biodiversity, ecosystem functioning and the provision of ecosystem services. *Biodivers. Conserv.* **2017**, *26*, 3005–3035. [CrossRef]
10. Matthies, B.D.; Kalliokoski, T.; Ekholm, T.; Hoen, H.F.; Valsta, L.T. Risk, Reward, and payments for ecosystem services: A portfolio approach to ecosystem services and forestland investment. *Ecosyst. Serv.* **2015**, *16*, 1–12. [CrossRef]
11. Martin-Ortega, J.; Dekker, T.; Ojea, E.; Lorenzo-Arribas, A. Dissecting price setting efficiency in Payments for Ecosystem Services: A meta-analysis of payments for watershed services in Latin America. *Ecosyst. Serv.* **2019**, *38*, 100961. [CrossRef]
12. Müller, A.; Olschewski, R.; Unterberger, C. The valuation of forest ecosystem services as a tool for management planning—A choice experiment. *J. Environ. Manag.* **2020**, *271*, 111008. [CrossRef]
13. Roesch-McNally, G.E.; Rabotyagov, S.S. Paying for forest ecosystem services: Voluntary versus mandatory payments. *Environ. Manag.* **2016**, *57*, 585–600. [CrossRef] [PubMed]
14. Rabotyagov, S.S.; Lin, S.J. Small forest landowner preferences for working forest conservation contract attributes: A case of Washington State, USA. *J. For. Econ.* **2013**, *19*, 307–330. [CrossRef]
15. Vedel, S.E.; Jacobsen, J.B.; Thorsen, B.J. Forest owners' willingness to accept contracts for ecosystem service provision is sensitive to additionality. *Ecol. Econ.* **2015**, *113*, 15–24. [CrossRef]
16. Leimona, B.; Carrasco, L.R. Auction winning, social dynamics and non-compliance in a payment for ecosystem services scheme in Indonesia. *Land Use Policy* **2017**, *63*, 632–644. [CrossRef]
17. Ferguson, I.; Levetan, L.; Crossman, N.; Lauren, B. Financial Mechanisms to Improve the Supply of Ecosystem Services from Privately-Owned Australian Native Forests. *Forests* **2016**, *7*, 34. [CrossRef]
18. Cook, D.C.; Kristensen, N.P.; Liu, S. Coordinated service provision in payment for ecosystem service schemes through adaptive governance. *Ecosyst. Serv.* **2016**, *19*, 103–108. [CrossRef]
19. Tóth, S.F.; Ettl, G.J.; Könnyű, N.; Rabotyagov, S.S.; Rogers, L.W.; Comnick, J.M. ECOSEL: Multi-Objective Optimization to Sell Forest Ecosystem Services. *For. Policy Econ.* **2013**, *35*, 73–82. [CrossRef]
20. Ninan, K.N.; Inoue, M. Valuing Forest Ecosystem Services: What We Know and What We Don't. *Ecol. Econ.* **2013**, *93*, 137–149. [CrossRef]
21. Roesch-McNally, G.E.; Rabotyagov, S.; Tyndall, J.C.; Ettl, G.; Tóth, S.F. Auctioning the Forest: A Qualitative Approach to Exploring Stakeholder Responses to Bidding on Forest Ecosystem Services. *Small Scale For.* **2016**, *15*, 321–333. [CrossRef]
22. Fujii, H.; Sato, M.; Managi, S. Decomposition Analysis of Forest Ecosystem Services Values. *Sustainability* **2017**, *9*, 687. [CrossRef]

23. McDermott, M.; Mahanty, S.; Schreckenberg, K. Examining Equity: A Multidimensional Framework for Assessing Equity in Payments for Ecosystem Services. *Environ. Sci. Policy* **2013**, *33*, 416–427. [CrossRef]
24. Ouyang, Z.; Hua, Z.; Yang, X.; Polasky, S.; Daily, G.C. Improvements in ecosystem services from investments in natural capital. *Science* **2016**, *352*, 1455–1459. [CrossRef] [PubMed]
25. Fiorini, A.O.; Mullally, C.; Swisher, M.; Putz, F.E. Forest cover effects of payments for ecosystem services: Evidence from an impact evaluation in Brazil. *Ecol. Econ.* **2020**, *169*, 106522.
26. Palomo, I. Climate Change Impacts on Ecosystem Services in High Mountain Areas: A Literature Review. *Mt. Res. Dev.* **2017**, *37*, 179–187. [CrossRef]
27. Rabotyagov, S.S.; Tóth, S.F.; Ettl, G.J. Testing the Design Variables of ECOSEL: A Market Mechanism for Forest Ecosystem Services. *For. Sci.* **2013**, *59*, 303–321. [CrossRef]
28. Lundberg, L.; Persson, U.M.; Alpizar, F.; Lindgren, K. Context Matters: Exploring the Cost-Effectiveness of Fixed Payments and Procurement Auctions for PES. *Ecol. Econ.* **2018**, *146*, 347–358. [CrossRef]
29. McGrath, F.L.; Carrasco, L.R.; Leimona, B. How auctions to allocate payments for ecosystem services contracts impact social equity. *Ecosyst. Serv.* **2017**, *25*, 44–55. [CrossRef]
30. Wunder, S.; Börner, J.; Ezzine-de-Blas, D.; Feder, S.; Pagiola, S. Payments for Environmental Services: Past Performance and Pending Potentials. *Annu. Rev. Resour. Econ.* **2020**, *12*, 209–234. [CrossRef]
31. Van Hecken, G.; Bastiaensen, J.; Windey, C. Towards a Power-Sensitive and Socially-Informed Analysis of Payments for Ecosystem Services (PES): Addressing the Gaps in the Current Debate. *Ecol. Econ.* **2015**, *120*, 117–125. [CrossRef]
32. Farley, J.; Costanza, R. Payments for ecosystem services: From local to global. *Ecol. Econ.* **2010**, *69*, 2060–2068. [CrossRef]
33. Salzman, J.; Bennett, G.; Carroll, N.; Goldstein, A.; Jenkins, M. The global status and trends of Payments for Ecosystem Services. *Nat. Sustain.* **2018**, *1*, 136–144. [CrossRef]
34. Etchart, N.; Freire, J.L.; Holland, M.B.; Jones, K.W.; Naughton-Treves, L. What Happens When the Money Runs out? Forest Outcomes and Equity Concerns Following Ecuador's Suspension of Conservation Payments. *World Dev.* **2020**, *136*, 105124. [CrossRef]
35. Phan, T.-H.D.; Brouwer, R.; Hoang, L.P.; Davidson, M.D. A comparative study of transaction costs of payments for forest ecosystem services in Vietnam. *For. Policy Econ.* **2017**, *80*, 141–149. [CrossRef]
36. Chib, S.; Greenberg, E. Analysis of Multivariate Probit Models. *Biometrika* **1998**, *85*, 347–361. [CrossRef]
37. Bingham, L.R.; Riccardo, R.D.; José, G.B. Ecosystem Services Auctions: The Last Decade of Research. *Forests* **2021**, *12*, 578. [CrossRef]
38. Persson, U.M.; Alpízar, F. Conditional cash transfers and payments for environmental services—A conceptual framework for explaining and judging differences in outcomes. *World Dev.* **2013**, *43*, 124–137. [CrossRef]
39. Brown, L.K.; Troutt, E.; Edwards, C.; Gray, B.; Wanjing, H. A uniform price auction for conservation easements in the Canadian prairies. *Environ. Resour. Econ.* **2011**, *50*, 49–60. [CrossRef]
40. Naime, J.; Mora, F.; Sánchez-Martínez, M.; Arreola, F.; Balvanera, P. Economic valuation of ecosystem services from secondary tropical forests: Trade-offs and implications for policy making. *For. Ecol. Manag.* **2020**, *473*, 118294. [CrossRef]
41. Oh, C.O.; Lee, S.; Kim, H.N. Economic Valuation of Conservation of Inholdings in Protected Areas for the Institution of Payments for Ecosystem Services. *Forests* **2019**, *10*, 1122. [CrossRef]
42. Richards, R.C.; Petrie, R.; Christ, J.B.; Ditt, E.; Kennedy, C.J. Farmer preferences for reforestation contracts in Brazil's Atlantic Forest. *For. Policy Econ.* **2020**, *118*, 102235. [CrossRef]
43. Ranjan, R. Protecting warming lakes through climate-adaptive PES mechanisms. *Ecol. Econ.* **2020**, *177*, 106782. [CrossRef]
44. Davis, R.K. Recreation Planning as An Economic Problem. *Nat. Resour. J.* **1963**, *3*, 239–249.
45. Lancaster, K.J. A New Approach to Consumer Theory. *J. Polit. Econ.* **1966**, *74*, 132–157. [CrossRef]
46. Cane, V.; Luce, R.D. Individual Choice Behavior: A Theoretical Analysis. *J. R. Stat. Soc. Ser. A (Gen.)* **1960**, *123*, 486–488. [CrossRef]
47. Mcfadden, D. Conditional Logit Analysis of Qualitative Choice Behavior. In *Frontiers of Econometrics*; Zarembka, P.E., Ed.; Academic Press: New York, NY, USA, 1973.
48. Louviere, J.; Hensher, D.A. On the design and analysis of simulated choice or allocation experiments in travel choice modeling. *Transp. Res. Rec.* **1982**, *890*, 11–17.
49. Louviere, J.J.; Woodworth, G. Design and Analysis of Simulated Consumer Choice or Allocation Experiments: An Approach Based on Aggregate Data. *J. Mark. Res.* **1983**, *20*, 350–367. [CrossRef]
50. Vickrey, W. Counter speculation, auctions, and competitive sealed tenders. *J. Financ.* **1961**, *16*, 8–37. [CrossRef]
51. Becker, G.M.; DeGroot, M.H.; Marschak, J. Measuring utility by a single-response sequential method. *Behav. Sci.* **1964**, *9*, 226–232. [CrossRef]
52. Shogren, J.F.; Margolis, M.; Koo, C.; List, J.A. A random nth price auction. *J. Econ. Behav. Organ.* **2001**, *46*, 409–421. [CrossRef]
53. Lusk, J.L.; Alexander, C.; Rousu, M.C. Designing Experimental Auctions for Marketing Research: The Effect of Values, Distributions and Mechanisms on Incentives for Truthful Bidding. *Rev. Mark. Sci.* **2011**, *5*, 1546–5616. [CrossRef]
54. Lewis, D.J.; Polasky, S. An auction mechanism for the optimal provision of ecosystem services under climate change. *J. Environ. Econ. Manag.* **2018**, *92*, 20–34. [CrossRef]
55. Sharma, B.P.; Cho, S.H.; Yu, T.E. Designing cost-efficient payments for forest-based carbon sequestration: An auction-based modeling approach. *For. Policy Econ.* **2019**, *104*, 182–194. [CrossRef]

Article

Valuing Recreational Services of the National Forest Parks Using a Tourist Satisfaction Method †

Nannan Kang [1,2], Erda Wang [2], Yang Yu [3,*] and Zenghui Duan [2,4]

1 China Center for Agricultural Policy (CCAP), School of Advanced Agricultural Sciences, Peking University, Beijing 100871, China; nnkang.ccap@pku.edu.cn
2 School of Economics and Management, Dalian University of Technology, Dalian 116024, China; edwang@dlut.edu.cn (E.W.); duanzenghui@dlou.edu.cn (Z.D.)
3 School of Economics and Management, Dalian Ocean University, Dalian 116024, China
4 Applied Technology College, Dalian Ocean University, Dalian 116300, China
* Correspondence: Yuyang770727@163.com; Tel.: +86-0411-84707090

† This paper is an expanded version of our working paper published in the 2020 3rd International Conference on Economic Management and Green Development (ICEMGD 20), Dalian University of Technology in 2019.

Abstract: Estimating the economic value of ecosystem services has become one of the most fertile areas in ecological economics. In this paper, we propose a novel method of using a tourist satisfaction model to evaluate the recreational services being embedded in forest ecosystems. We establish a functional relationship between tourist satisfaction and recreational attributes based on the survey data of China National Forest Parks. The results indicate that each recreational attribute considered enables the generation of a significant amount of tourism welfare for tourists, whereas tourist congestion was found to be a negative contributor to tourists' satisfaction. Reducing congestion from the current level is the most valued recreational attribute for tourists, and the willingness to pay for it is as high as CNY 623.18 (USD 92.29) per visitor per trip. Additionally, local and nonlocal tourists display a divergent degree of preference for the recreational attributes and their levels of willingness to pay.

Keywords: economic value; recreational services; tourist satisfaction; national forest parks

Citation: Kang, N.; Wang, E.; Yu, Y.; Duan, Z. Valuing Recreational Services of the National Forest Parks Using a Tourist Satisfaction Method. *Forests* **2021**, *12*, 1688. https://doi.org/10.3390/f12121688

Academic Editors: Elisabetta Salvatori and Giacomo Pallante

Received: 16 November 2021
Accepted: 29 November 2021
Published: 2 December 2021

1. Introduction

Forest ecosystem services and the natural capital stocks that produce them are considered an important pillar of human life satisfaction [1]. Although there is growing recognition of the need for their conservation, ecosystems continue to be lost through the world [2]. A key factor behind this degradation is that most ecosystem services have the characteristics of public or quasi-public goods, causing difficulties for the users to be able to accurately identify their true values. Consequently, people can profit from the non-consumptive trade and no incentive exists to pay to maintain them. Therefore, significant efforts are needed to create a market in which ecosystem services can be found, defined, and calculated.

In the past two decades, a growing body of research in ecological economics has concerned satisfaction data and investigated the determinants of satisfaction at the individual level. As respondent's self-reported satisfaction data for ecosystem services can serve as a metric for a personal utility level, valuing ecosystem services directly in satisfaction terms have sparked a strong interest in the field of non-market valuation [3–5].

The basic idea of the satisfaction-based method is simple. This method uses survey data on self-reported satisfaction as an empirical approximation to "the experienced utility" or individual welfare, and models individual satisfaction as a function of their incomes or costs, ecosystem services they used, and other control variables. The estimated parameters are then used to calculate the average marginal rate of substitution between the level of income or cost and ecosystem service. As such, the individual's marginal willingness to pay (WTP) for the service in question can be revealed [5].

To some extent, the satisfaction-based method exhibits certain desirable features or advantages over traditional tools (e.g., the stated preference method and the revealed preference method). For example, there is no need to directly ask about people's willingness to pay in order for the public goods to be valued; instead, it simply requires an individual to rank his or her life satisfaction [6]. As such, it can greatly mitigate the survey burden for both research practitioners and survey participants. Furthermore, the satisfaction survey can effectively avoid strategic bias, which is a common problem in the contingent valuation survey process. In addition, this method can also suppress the model estimation bias that is potentially caused by the strong assumptions of a rational economic actor and the complete market equilibrium condition [7].

Welsch (2002) was the first researcher who used the cross-section life satisfaction data gathered from 54 countries [8]. Since then, there has been a burgeoning number of studies focused on valuing air quality in different countries and geographical locations, in addition to various economic development stages globally, all based on life satisfaction data [6,9,10]. In recent years, the life satisfaction data has gradually gained its foothold in other fields of ecological economics, including valuing aesthetic services [11,12], biodiversity services [13], and cultural goods [3], in addition to the negative externalities of wind turbines [14]. To the best of our knowledge, such a study focused on recreational ecosystem services using tourist satisfaction (TS) data has not been carried out to date.

With respect to recreational services of ecosystems, some studies have demonstrated that there is a close interrelationship between ecosystems and value received by tourists [15,16]. The tourist experiences at a destination are likely to result in positive memories, emotional attachment, and, eventually, tourist satisfaction [17,18]. Furthermore, tourist satisfaction accrued from unforgettable experiences and emotional attachment signifies recreational value for tourists. Barbara (1997) suggested if one's actual experience on a recreational site is better than his or her expectation, then his or her assessment on the site should be considered as being 'satisfactory and valuable' [19]. Similarly, Williams et al. (2003) stated that the value of natural space can be derived from people's satisfaction or their emotional attachment to the space [20]. Therefore, based on these research analyses and insightful arguments, we can extend the satisfaction-based method to the area of valuing recreational services of national forest parks (NFPs). Specifically, we can build a micro-econometric tourist satisfaction function using travel cost, the site attributes, and several other covariates as arguments. Then, the estimated model parameters can be used in computing the recreational attribute's economic values.

This study makes the following contributions to the current literature: (i) Exploring the underlying theory of tourist satisfaction measurement and its connection with tourist welfare. Our proposed theoretical framework can be used in assessing the economic value of forest recreational services. In this study, we discuss the advantages of the TS approach compared to the conventional methods. (ii) Regarding tourism marketing, it is meaningful to examine the differences in economic value that are attributable to the different tourist groups, including local tourists and nonlocal park tourists. (iii) It is expected that the analytical results will have implications for park management regarding park budget allocation, level of admission fees charged, over-crowding control, priority order of park management tasks, and park marketing strategies.

The remainder of the paper is organized as follows. Section 2 proposes an empirical model framework, and outlines the questionnaire design and data collection process. Section 3 addresses the model results, and is followed by Section 4, which presents the study summary and conclusions.

2. Materials and Methods

2.1. Methodological Framework

2.1.1. Utility Correlates to Economic Value

The economic concept of value employed here originates from neoclassical welfare economics. The basic premises of welfare economics are that the purpose of economics is to increase the well-being of the individuals who make up the society, and that each individual is the best judge of how well off he or she is in a given situation. Each individual's welfare depends not only on that individual's consumption of private goods, and of goods and services produced by the government, but also on the quantities and qualities of nonmarket goods and service flows received by each individual from the resource–environment system [21]. In the context of tourism economics, each tourist's utility depends not only on his or her consumption of private goods, such as transportation, goods, and lodging services, but also on the quantities and qualities of non-market recreational service flows each tourist receives from the national park's ecological system; for example, health, visual amenities, and opportunities for outdoor recreation. Based on their utility-maximizing behavior, tourists' valuation of recreational services is revealed by their consumption choices [22].

For a tourist who makes consumptive decisions concerning bundles of recreational goods and services, his/her utility, which represents preferences for various recreational attributes and tourism consumption, is given by the property of substitutability among the market and non-market goods which make up the bundles. By substitutability, we mean that if the quantity of one service in the consumed bundle is reduced, it is possible to increase the quantity of other services so as to leave the tourist no worse off as a result of the changes in the bundle. This substitutability is known to represent market demand as if it is the outcome of a decision by a rational consumer [23]. The property of substitutability measured by the marginal rate of substitution (MRS) is at the core of the economist's concept of value because substitutability establishes trade-off ratios between pairs of services that matter to the tourists.

The value measures based on substitutability can be expressed either in terms of willingness to pay (WTP) or willingness to accept compensation (WTA). WTP and WTA measures can be defined in terms of any other service that the tourist is willing to substitute for the service being valued. In valuing the NFP services, we use travel cost as the numeraire so that the trade-off ratios between currency term and a specific park characteristic can be expressed. The WTP is the maximum sum of money that an individual would be willing to pay rather than do without an increase in some service, such as the amount of rubbish reduction in a NFP site [24]. More formerly, a tourist's utility function can be formulated as the following:

$$u = f(r, t, \theta)$$

where u denotes the tourist's utility, and r, t, and θ denote the level of recreational attributes, amount of the travel cost, and tourist's personal traits, respectively.

2.1.2. Utility and Tourist Satisfaction

The study of people's satisfaction with their life or the service quality they experience has long been one of the mainstays in psychology. Only since the "Easterlin paradox" was proposed has this psychological research been connected to economics [25]. The key proposition associated with the theoretical framework is that an individual's true utility is an implicit variable and unobservable, which leads to a controversy about whether self-stated satisfaction is a reliable and valid proxy for the tourist's utility. In general, tourists pursue individual welfare based on some stable evaluation metrics. For a measure of reported satisfaction to serve as a proxy for individual welfare, an important assumption is necessary: the standards underlying people's judgment are those the individual would like to pursue in realizing his/her ideal of a pleasant travel experience.

Tourist satisfaction refers to the perceived discrepancy between prior expectation and perceived performance after consumption. The extent to which tourist satisfaction is identified depends on whether the evaluation standards fit the tourist's judgments about their travel experience. When experiences compared to expectations result in feelings of gratification, the tourist is satisfied, and vice versa [26]. For tourists, the general evaluation of the satisfaction and pleasure associated with their travel may be an appropriate standard to capture judgments about their welfare. Therefore, like most researchers who use TS as a valid proxy for respondent's utility, we employ tourist satisfaction as a measure of tourist's utility [11,27].

2.1.3. Empirical Model

A typical tourist satisfaction model depicts the relationship between the parks' recreational services, tourist's travel cost, and other control variables. Concretely, these can be expressed in Equation (1) below:

$$U_{ij} = \alpha_0 + \beta_i' RA_{ij} + \mu_i' OA_{ij} + \gamma_i' C_{ij} + \delta_i' S_{ij} + \varepsilon_{ij} \tag{1}$$

where U_{ij} is the true but unobservable indirect utility of tourist i as he or she travels to the park j; $RA_{ij} = \left(RA_{ij1}, RA_{ij2}, \ldots RA_{ijk}\right)'$ is a vector representing the recreational attributes of the park j; and $\beta_i' = (\beta_{i1}, \beta_{i2}, \ldots \beta_{ik})'$ is a vector of the park attribute coefficients. Similarly, $OA_{ij} = \left(OA_{ij1}, OA_{ij2}, \ldots OA_{ijk}\right)'$ is a vector of the control variables associated with the park j, which may involve factors such as park land size, temperature, and rainfall; C_{ij} represents a set of travel costs; S_{ij} is a vector of tourists' socio-economic and demographic characteristics; and ε_{ij} is a random error.

As discussed above, the key step of measuring the park service's value is to correctly define and formulate the marginal rate of substitution between the recreational attribute and the tourists' travel cost. To do so, we must first calculate the marginal utility of the park attribute and the marginal utility of the tourists' travel cost; then, by taking the ratio between the two, we can derive the monetary value of the attributes. The MRS can then be directly derived from Equation (1), as presented in Equation (2):

$$MRS = \frac{\Delta Cos\ t}{\Delta\ Recreation\ attribute} = -\frac{\partial U_{ij} / \partial RA_{ij}}{\partial U_{ij} / \partial C_{ij}} = -\frac{\hat{\beta}}{\hat{\gamma}} \Leftrightarrow WTP \tag{2}$$

2.2. Questionnaire Design and Data Collection

2.2.1. Questionnaire Design

In this study, we developed two survey questionnaires: one was used to capture data on the quality of recreational attributes of the NFPs, and the other was used to collect data on the satisfaction of tourists and their socio-economic characteristics.

1. Rreational attributes of NFPs: The tourist satisfaction or tourist welfare primarily depends on a range of recreational attributes of the NFPs. To correctly identify these attributes, we conducted an extensive literature review, which enabled us to identify the six core park attributes of the level of tourist satisfaction, including the park natural resources [28–30], accessibility [31,32], infrastructure that relates to tourists' basic needs [31,33], and park management [29]. Table 1 shows the details of these attributes in addition to several control variables considered in this study.
2. Measurement of tourist satisfaction: The level of tourist satisfaction was measured using the China tourist satisfaction index, which was developed by "The China tourist satisfaction evaluation system" in 2012 and is administrated by the China National Tourism Administration. The tourist satisfaction index is characterized by five categories: flat satisfaction, loyalty, demand, expectation, and recommend intention. Each category is measured using a 5-point Likert scale from 'Completely dissatisfied' to 'Completely satisfied'. The validity tests of this satisfaction index are presented in Table 2.

3. Travel cost: This covers all the travel expenses occurred during the travel process, from a tourist's home origin to the park destination. The basic calculation formula is given by Equation (3):

$$Cos\ t = C_1/N_1 + C_2 + C_3 \quad (3)$$

where C_1 represents the expenses paid before a tourist arrives at the entrance point of the forest park, which covers the costs paid for transportation, lodging, food and beverage, and other. N_1 represents the number of sites visited by the tourist, which is used in allocating the costs to a specific park site to avoid the problem of possible double counting. C_2 is the cost spent at the park site, including admission fees, amusement, food and beverages, accommodation, and souvenirs. C_3 denotes the time opportunity cost, which is computed using Equation (4):

$$1/3 \times (T_1 + 2 \times T_2)/(Y \div 12 \div 23 \div 24) \quad (4)$$

where T_1 is the length of time spent touring the park site (unit: h); T_2. is the time taken on one-way travel from the home to the park (unit: h); Y is the household income; 12 means the number of months during which a tourist works in a year, excluding the legal weekends; the work time per month is calculated as 23 days.

4. Socio-economic characteristics. These include age, sex, education, marital status, household income, and whether a respondent is a repeat tourist or is visiting from areas outside of the park [34].

2.2.2. Data Collection

The data collection involved two aspects: one related to the physical conditions of the recreational services, was provided by the park management personnel, and included the natural environment, the park history, and the park operations. The other regarded the tourist's personal trait information, such as travel satisfaction, money spent during the tour, and demographic data. Before the formal survey activity was undertaken, a statistical pretest was administrated to ensure the survey questions were understandable and meaningful to the respondents. The pretests were based on data gathered from 4 park managers and 30 tourists at Yin-Shi-Tan NFP and Xi-Jiao NFP, respectively, in Western Dalian, China, and were implemented during 10–15 April 2017. The pretest outcomes were used to check and upgrade the survey questionnaires.

As shown in Figure 1, the study sample covered 22 NFPs across over 14 provinces in China, and thus represented a wide range of conditions of the NFPs in China with regards to the park grade rankings, land size, length of establishment, rates of forest vegetation, etc. (see Table S1 for detail). The whole survey process lasted for about 7 months, from early April to December in 2017. Due to the constraints of both time and budget, it was not possible for the researchers to travel to every sampled park site to carry out field survey activities. Thus, most survey activities were delegated to the parks' management personnel, with the mutually agreed service fees paid to the clients. Thus, the tourist surveys were executed by the park managers on behalf of our research team members. On average, 200–300 tourists were sampled from each selected park site. A total of 4531 valid questionnaires were collected.

Table 1. The definition of the recreational attributes and the type of variable specified.

	Attribute	Attribute Description	Type of Variable	Variable Name
Recreational attributes	Rate of vegetation coverage	The vegetation coverage rate <60% The vegetation coverage rate 60–85% The vegetation coverage rate >85%	Dummy	Forest * $Forest^{+}$ $Forest^{++}$
	Quantity of rubbish	No. garbage can distributed per 200 m: 1; No. pieces in a visible scope: >10 No. garbage can distributed per 200 m: 2; No. pieces in a visible scope: 3–10 No. garbage can distributed per 200 m: 4; No. pieces in a visible scope: <3	Dummy	$Garbage^{-}$ Garbage * $Garbage^{+}$
	Traffic condition	Less convenient, Time spent from city to park: >180 min Partially convenient, Time spent from city to park: 60–180 min Well convenient, Time needed from city to park: <60 min	Dummy	Traffic * $Traffic^{+}$ $Traffic^{++}$
	Congestion	No. people observed in a visible scope(per 100 m^2): >60 No. people observed in a visible scope(per 100 m^2): 50 No. people observed in a visible scope(per 100 m^2): 35 No. people observed in a visible scope(per 100 m^2): 20 No. people observed in a visible scope(per 100 m^2): <10	Dummy	$Congestion^{--}$ $Congestion^{-}$ Congestion * $Congestion^{+}$ $Congestion^{++}$
	Support facility	Elements including eco-lavatory, wood path, parking lot, service center, and special eateries and shops, each item earns 1 point (Excellent = 5, Good = 4, Medium = 3, Average = 2,Inferior = 1).	Continuous	Support
	Recreation facility	Including 5 factors such as playground, song and dance, tourism guide service, sightseeing vehicle, as well as a channel of the official information. With each item earning 1 point. (Excellent = 5, Good = 4, Medium = 3, Average = 2, Inferior = 1).	Continuous	Recreation
Control variable	Area	National Forest Parks (hm^2)	Continuous	Area
	Temperature	Average monthly temperature (°C)	Continuous	Temperature
	$Temperature^{2}$	Average monthly temperature squared (°C)	Continuous	$Temperature^{2}$
	Rainfall	Average monthly precipitation (mm)	Continuous	Rainfall
	Air	Aero-anion concentration monitored every quarter (10,000/cm^3)	Continuous	Air
	Humanity	Number of cultural attractions with historical and cultural heritage open to the public	Continuous	Humanity

Note: Variable with "*" stands for a baseline status.

Table 2. Measurement of tourist satisfaction and its validity tests.

Tourist Satisfaction Scale		Score
In general, I am satisfied with this park visit. This visit meets my expectation. The overall travel meets my expectation. If there is an opportunity, I would like to revisit this site. I would like to recommend the park to my friends and relatives.		5-pt Likert Scale: 1 'completely agree' 5 'completely disagree'
Number of items		5
Cronbach's alpha coefficient		0.854
Kaiser–Meyer–Olkin Measure		0.893
Bartlett's Test of Sphericity	Approx. Chi-Square	11,990.865
	df.	21
	Sig.	0.000

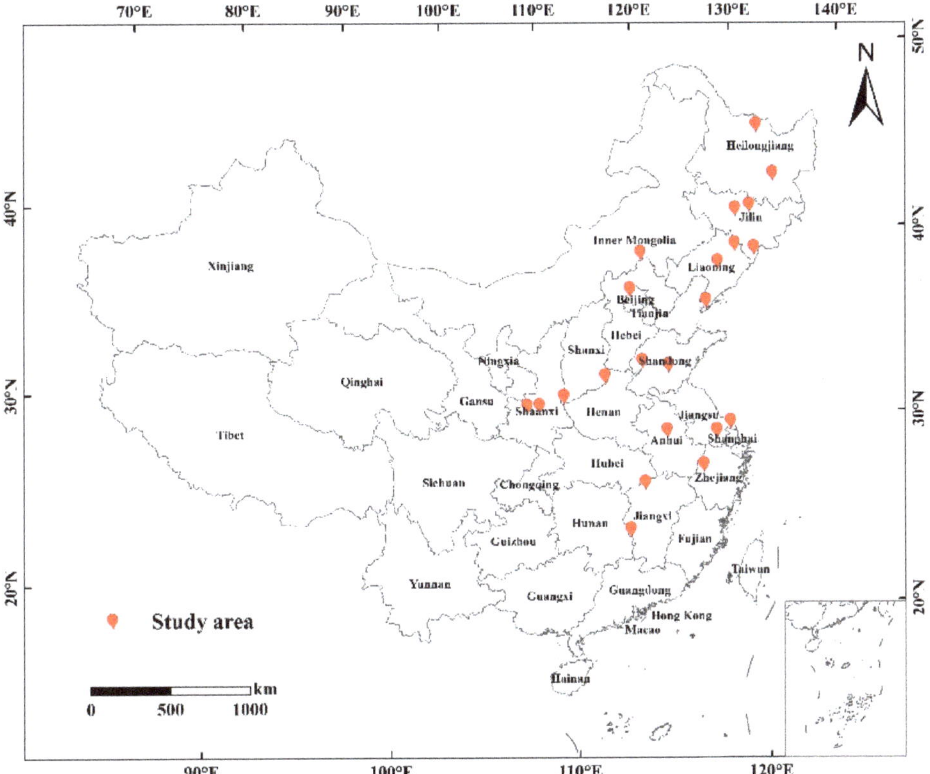

Figure 1. Geographical distribution of 22 National Forest parks.

3. Results and Discussion

As mentioned in the literature review, the tourist satisfaction function can be properly estimated using an ordered probit model [9,35]. If the satisfaction function is characterized as being cardinal, then the satisfaction function can be estimated via the ordinary least squares (OLS) model [27]. However, numerous studies have shown that treating satisfaction as either an ordinal or cardinal variable makes little difference to the estimated model results [36]. Therefore, here we chose to use both ordinary least squares and ordered probit models for estimating model parameters. This analogy is very similar to the life satisfaction measurement commonly used by non-market valuation researchers [37].

Based on the origins of the tourists, we subdivided the whole survey dataset into two subsamples: a local tourist group and a nonlocal tourist group. A local tourist was defined as a tourist who resided in the same city as the location of the NFP; and a nonlocal tourist was defined as one whose residential location differed from that of the NFP they visited. The detailed descriptive statistics for each sample dataset are presented in Table S2.

3.1. Model Results for the Whole Sample

To better understand the underlying effects of the NFP recreational attributes on tourist satisfaction, we begin by estimating the models defined by Equation (1) using the entire dataset; the results are presented in Table 3. Columns 4–5 in Table 3 show the results estimated from Equation (1), based on the ordered probit model in which tourist satisfaction is considered as the latent variable. It can be seen clearly that both the OLS and the ordered probit models have a high goodness of fit. Furthermore, the two models generated very similar results in terms of the magnitudes of the model coefficients and their statistical performance.

Table 3. The model results based on the whole dataset.

Variable	Linear Form		Ordered Probit Model	
	Coef.	S.E.	Coef.	S.E.
Cost	-3.66×10^{-4} ***	6.74×10^{-5}	-4.33×10^{-4} ***	8.66×10^{-5}
Forest^{+}	0.066 **	0.033	0.116 ***	0.043
Forest^{++}	0.120 ***	0.037	0.200 ***	0.048
Traffic^{+}	0.037	0.035	0.069	0.045
Traffic^{++}	0.041	0.026	0.048	0.034
Garbage^{-}	−0.103 ***	0.035	−0.145 ***	0.045
Garbage^{+}	0.131 ***	0.033	0.142 ***	0.043
Congestion^{--}	−0.166 ***	0.032	−0.174 ***	0.042
Congestion^{-}	−0.092 ***	0.034	−0.127 ***	0.043
Congestion^{+}	0.228 ***	0.035	0.281 ***	0.046
Congestion^{++}	−0.091 **	0.046	−0.182	0.059
Support facility	0.040 ***	0.016	0.080 ***	0.021
Recreation facility	0.045 ***	0.017	0.093 ***	0.021
Humanity	−0.001	0.001	−0.001	0.001
Air	0.045 ***	0.005	0.062 ***	0.007
Area	−0.0001 **	5.75×10^{-5}	-0.28×10^{-4} ***	7.83×10^{-5}
Temperature	−0.001	0.001	−0.001	0.001
Temperature$^{\hat{}2}$	−0.001 ***	0.000	−0.002 *	0.000
Rainfall	−0.007	0.007	−0.012	0.009
Revisit or not	−0.051 *	0.026	−0.071	0.034
Age	−0.013 *	0.007	−0.014	0.010
Age$^{\hat{}2}$	1.56×10^{-4} *	0.000	1.75×10^{-4}	1.05×10^{-4}
Gender	0.016	0.026	0.027	0.033
Education	−0.074 *	0.016	−0.099 ***	0.021
Marriage	0.015	0.020	0.019	0.026
HH income	0.000	0.009	−0.002	0.011
Residence	−0.267 ***	0.043	−0.278 ***	0.055
Cons	3.438	0.178	-	-
Number of observation	4531		4531	
R^2	0.1029		0.0495	
Log-likelihood			−5218.121	

Note: *** $p < 0.01$, ** $p < 0.05$, * $p < 0.1$.

Regarding the recreational services, there is a significant positive relationship between the two attributes ('Rate of vegetation coverage' and 'Quantity of rubbish') and tourist satisfaction. However, the attribute of 'Congestion' seems to impose a non-linear effect on tourist satisfaction because, as the level of congestion reaches a rate of 'more than 35 people/100 m^2', the crowding imposes a statistically negative effect on tourist satisfaction. However, as the level of congestion decreases to a level of '20 people/100 m^2', the crowding is shown to make a positive contribution to tourist satisfaction. Interestingly, the negative effect reoccurs as the congestion level reaches 'Less than 10 people/100 m^2'. This may reflect the fact that tourists exhibit a dual preference regarding the congestive conditions in the park, such that they neither appreciate over-congestion nor enjoy an environment at the site that is too isolated. Regarding the attribute of support facilities at the park sites, tourists prefer more convenient 'support facilities' provided in the parks. Nevertheless, this notion does not seem to be supported by the statistical test results because 'Convenience of transportation' does not significantly affect tourist satisfaction.

Among the considered tourists' traits, 'age', 'education' and 'origins' all affect TS significantly. However, 'age' exhibits a U-shaped relationship with TS in such a way that the level of satisfaction reaches the lowest as tourists attain middle age. As shown by Jarvis et al. (2016), tourists with a lower level of education tend to be more satisfied than those who received more advanced education [34]. Moreover, the origin of a tourist has a strong influence on his or her travel satisfaction ($\alpha = 0.01$). However, nonlocal tourists tend to be more likely to report a better level of satisfaction than do the local tourists. This may be attributable to the fact that tourists who reside in areas near the NFPs are used to or more familiar with the park environment; thus, to some extent, they may have a lesser feeling of freshness and are less sensitive to the level of satisfaction.

3.2. Valuing the National Forest Parks' Recreational Service

As discussed previously, the monetary value of the NFP's service is estimated using the marginal rate of substitution (MRS), i.e., by computing the ratios between the estimated recreational attribute coefficients and the tourist's travel cost. In order to account for the non-linear relationship between the variation in each attribute and the travel cost, we computed the MRS based on alternative combinations of travel costs and recreational attributes. Table 4 presents the mean WTP results derived from using the whole dataset. It is clear that the WTP results vary by model. However they do not seem to differ greatly, which indicates a high degree of robustness of the estimated model results.

Table 4. The economic values accrued to the changes in the park characteristics.

	Linear Form		Ordered Probit Model	
	WTP	Proportion	WTP	Proportion
Forest *	-	-	-	-
Forest^{+}	180.921	0.62	270.394	0.93
Forest^{++}	327.263	1.13	468.133	1.61
Traffic *	-	-	-	-
Traffic^{+}	100.846	0.35	162.410	0.56
Traffic^{++}	110.793	0.38	112.289	0.39
Garbage^{-}	−281.882	−0.97	−339.634	−1.17
Garbage *	-	-	-	-
Garbage^{+}	356.450	1.23	333.010	1.15
Congestion^{--}	−454.590	−1.56	−407.551	−1.40
Congestion^{-}	−251.720	−0.87	−297.075	−1.02
Congestion *	-	-	-	-
Congestion^{+}	623.184	2.14	655.801	2.26
Congestion^{++}	−249.647	−0.86	−426.315	−1.46
Support facility	109.289	0.38	184.988	0.63
Recreation facility	122.951	0.42	214.781	0.74

Note: 1 USD = 6.752 CNY in 2017. Proportion = WTP/mean travel cost. The mean value of travel cost in the whole sample is CNY 290.63 (USD 43.04). Variables with "*" represent a baseline status.

In the discussion below, we use the results generated from Model 1 to address the attribute values. This is simply because Model 1 has the best goodness of fit among the three models. It is interesting to see that, among all the recreational attributes, tourists attach the greatest importance to the attribute of 'Congestion^{+}' which is evidenced by the highest value of the WTP (CNY 623.18, i.e., USD 92.29), followed by the WTP for the attribute of vegetation coverage (CNY 327.26, i.e., USD 48.54). This indicates that tourists have a strong desire to experience a less crowded environment and better coverage of green land in the forest parks. However, as the congestion level changes from the status quo to the worst level ('Congestion^{--}'), the WTP falls to CNY 454.59 (USD 67.33). Similarly, the effect of changing the 'amount of rubbish' from the status quo to 'Garbage^{-}' results in an economic loss of CNY 281.88 (USD 41.75). Finally, the average WTP for a one-level increment in 'Support facility' and 'Recreation facility' is CNY 109.29 (USD 16.19) and CNY 122.95 (USD 18.21), respectively.

In contrast to Chinese tourists, who pay the greatest attention to the attribute of 'Congestion^{+}', Penn et al. (2016) found 'little congestion' was relatively less important in Hawaii beach, and tourists are only willing to pay USD 6.37 for ideal crowding conditions [38]. In China, the exploitation of forest tourism resources is far greater than that of ocean tourism. From April to November each year, i.e., when we collected the data in this study, multiple millions of tourists travel to NFPs, causing overcrowding in the parks. Therefore, tourists are more sensitive to changes in the congestion level in forest parks and ultimately show a higher willingness to pay.

3.3. Model Results for the Two Sub-Samples

To better examine the difference in the tourists' WTP for various recreational attributes between local and nonlocal tourist groups, we conducted a separate analysis on each group using the two types of models, i.e., linear and ordered probit models. The two columns listed under each model in Table 5 represent the estimated results developed from each sub-sample dataset.

As expected, the two tourist groups exhibit different attribute preferences and economic values. Regarding the nonlocal group of tourists, their satisfaction displays an inverted U-shaped relationship with the travel costs, such that the level of satisfaction reaches the highest level as the per person's travel expense reaches CNY 108.93 (USD 16.13). In comparison, the local group of tourists is subject to fewer cost constraints, so their satisfaction is not significantly affected by this factor. Regarding parks' recreational attributes, both 'Congestion' and 'Support facility' are statistically significant for the two groups of tourists. However, they differ in terms of the attribute of 'Convenience of traffic', such that the coefficient of the 'Best' level of traffic variable indicates a statistically significant positive effect for the local tourist group, whereas the effect is not significant for the nonlocal group. This may be because, relative to the nonlocal tourists, the local park tourists are more sensitive to the park conditions, such as whether it is more convenient for them to use the park facilities. In contrast, because they experienced a long journey from home to the NFP site, the nonlocal tourists tend to be less sensitive to the traffic conditions in the areas surrounding the park. Instead, the nonlocal tourists paid more attention to the parks' environmental quality, as reflected by 'Quantity of rubbish', and the diversity of recreational activities provided on the site (i.e., 'Recreation facility'). Hence, these attributes exhibit a statistically significant relationship with nonlocal tourists' satisfaction.

Table 6 presents the mean values of the WTP calculated from Equation (1) of the OLS and ordered probit models using the sub-datasets. In addition, we also carried out 1000 simulations using the parametric bootstrapping method to generate robust or stable estimated WTP outcomes [39]. To be consistent with the estimates using the complete dataset, we ran the models separately using the sub-datasets. It was found that using the sub-datasets yielded a higher WTP for the park attributes of 'Congestion' and 'Rate of vegetation coverage', such that the nonlocal tourists would be willing to pay CNY 556.03 (USD 82.35) as a tradeoff for a lower level of congestion in the parks; this is CNY 438.56 (USD 64.95) more than the local tourist's WTP. These results are similar to those of Lindberg and Veisten (2012), and indicate that the economic values that potentially accrue to the recreational attributes are different between local tourists and nonlocal tourists [40].

Table 5. Comparison between the two groups of tourists.

	Linear Form				Ordered Probit Model			
	Local Tourists		Nonlocal Tourists		Local Tourists		NonLocal Tourists	
	Coef.	S.E.	Coef.	S.E.	Coef.	S.E.	Coef.	S.E.
Cost	-2.54×10^{-3} ***	5.53×10^{-4}	$-3.23 \times^{4}$ ***	6.56×10^{-5}	-3.72×10^{-3} ***	6.43×10^{-4}	-3.93×10^{-4} ***	8.82×10^{-5}
Forest^{+}	0.068	0.075	0.067 *	0.037	0.105 *	0.087	0.101 **	0.050
Forest^{++}	0.152 **	0.082	0.112 ***	0.042	0.208 ***	0.097	0.133 ***	0.057
Traffic^{+}	0.020	0.075	0.004	0.040	0.083	0.087	0.013	0.054
Traffic^{++}	0.129 **	0.057	0.007	0.030	0.161 ***	0.066	0.003	0.040
Garbage^{-}	−0.111	0.074	−0.119	0.041	−0.157 *	0.086	−0.136 ***	0.055
Garbage^{+}	0.128 *	0.075	0.137	0.037	0.208 ***	0.090	0.131 ***	0.051
Congestion^{--}	−0.242 ***	0.089	−0.126 ***	0.038	−0.360 ***	0.104	−0.111 **	0.052
Congestion^{-}	−0.142 **	0.073	−0.096 **	0.045	−0.115 *	0.086	−0.166 ***	0.060
Congestion^{+}	0.298 ***	0.081	0.179 ***	0.044	0.303 ***	0.096	0.216 ***	0.061
Congestion^{++}	−0.148	0.101	−0.059	0.053	−0.107	0.118	−0.043	0.071
Support facility	0.131 ***	0.033	0.073 **	0.019	0.168 ***	0.038	0.103 ***	0.026
Recreation facility	0.051	0.038	0.047 ***	0.019	0.057	0.045	0.063 ***	0.025
Humanity	0.004 **	0.002	−0.002 **	0.001	0.003 **	0.002	−0.003 ***	0.001
Air	0.054 ***	0.012	0.037 ***	0.006	0.063 ***	0.014	0.051 ***	0.008
Area	2.98×10^{-4}	2.23×10^{-4}	-1.48×10^{4} *	0.60×10^{-4}	1.50×10^{-4}	2.62×10^{-4}	-2.97×10^{-4}	2.62×10^{-4}
Temperature	0.020	0.022	−0.001	0.001	0.005	0.026	−0.001	0.001
Temperature^2	−0.003 ***	0.001	−0.001 ***	0.000	−0.003 **	0.001	−0.002 ***	0.003
Rainfall	0.002	0.015	−0.009	0.007	−0.001	0.018	−0.014	0.010
Revisit or not	−0.064	0.056	−0.045	0.030	−0.081	0.065	−0.067	0.040
Age	−0.025	0.015	−0.002	0.008	−0.025	0.018	−0.002	0.011
Age^2	2.33×10^{-4}	1.78×10^{-4}	5.86×10^{-5}	0.91×10^{-4}	2.17×10^{-4}	2.08×10^{-4}	0.79×10^{-4}	1.25×10^{-4}
Gender	0.007	0.055	0.018	0.029	0.003	0.064	0.035	0.039
Education	−0.019	0.034	−0.095 ***	0.018	−0.024	0.040	−0.127 ***	0.024
Marriage	0.034	0.041	0.000	0.023	0.037	0.048	0.003	0.031
HH income	0.002	0.018	−0.002	0.010	0.001	0.021	−0.006	0.013
Cons	3.187	0.374	3.324	0.202	-	-	-	-
No. of observation	1205		3326		1205		3326	
R^2	0.1341		0.1039		0.0526		0.0539	
Log-likelihood					−1472.593		−3704.635	

Note: *** $p < 0.01$, ** $p < 0.05$, * $p < 0.1$.

Table 6. The economic values accrued to the recreational attributes based on different tourist groups.

	Linear Form				Ordered Probit Model			
	Local Tourist		Nonlocal Tourist		Local Tourist		Nonlocal Tourist	
	WTP	Proportion	WTP	Proportion	WTP	Proportion	WTP	Proportion
Forest *	-	-	-	-	-	-	-	-
Forest^{+}	26.58	0.33	208.11	0.57	28.38	0.36	255.23	0.70
Forest^{++}	59.73	0.75	348.00	0.95	56.05	0.70	336.49	0.92
Traffic *	-	-	-	-	-	-	-	-
Traffic^{+}	8.04	0.10	12.67	0.03	22.39	0.28	33.19	−0.09
Traffic^{++}	50.92	0.64	23.14	0.06	43.30	0.54	6.55	0.02
Garbage^{-}	−43.63	−0.55	−369.05	−1.01	−42.27	−0.53	−346.47	−0.94
Garbage *	-	-	-	-	-	-	-	-
Garbage^{+}	50.29	0.63	424.16	1.16	56.05	0.70	332.63	0.91
Congestion^{--}	−95.09	−1.20	−389.58	−1.06	−96.77	−1.22	−281.65	−0.77
Congestion^{-}	−55.95	−0.70	−297.15	−0.81	−47.15	−0.59	−421.73	−1.15
Congestion *	-	-	-	-	-	-	-	-
Congestion^{+}	117.47	1.48	556.03	1.51	81.58	1.03	549.46	1.50
Congestion^{++}	−58.43	−0.73	−182.66	−0.50	−28.76	−0.36	−110.33	−0.30
Support facility	51.55	0.65	226.12	0.62	45.30	0.57	260.83	0.71
Recreation facility	20.08	0.24	145.43	0.40	15.47	0.19	160.09	0.44
Total	138.05	1.73	813.38	2.22	106.93	1.34	829.32	2.26

Note: 1 USD = 6.752 CNY in 2017. Variables with "*" represent a baseline status.

For example, regarding the attribute of 'Congestion^{+}', the distribution of tourists' WTP for the related attributes is depicted in Figure 2 based on the results presented in Table 6. The x-axis represents the WTP, and the y-axis represents the probability distribution of the WTP. The red dotted line shows the WTP distribution for the local tourist group and the blue solid line shows that for the nonlocal tourist group. Obviously, the nonlocal tourist group exhibits a much higher WTP than the local tourist group. Nonetheless, the latter displays a wider dispersion of distribution. A possible reason for this could be that the higher level of WTP for these important NFP attributes indicates the nonlocal tourists, in general, attach the greatest importance to the parks they chose to visit. It is also likely that these nonlocal tourists have a relatively higher expectation than the local tourists to be motivated to make the decision to travel to the NFPs. Because the expectation and actual perception do not considerably differ, tourists would be willing to pay more money to the park sites they visited. The descriptive statistics show that most tourists expressed a higher level of tourist satisfaction, which seems to provide the underlying reason why the nonlocal tourist group exhibits a higher level of WTP than the local tourists.

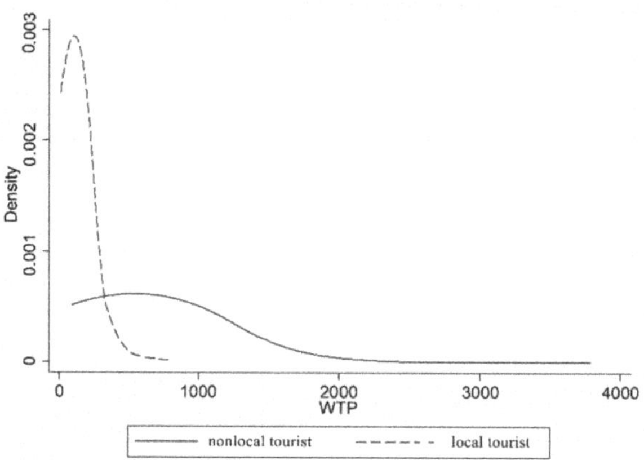

Figure 2. The WTP's distribution with the two different datasets, taking 'Congestion$^+$' as an example; 1 USD = 6.752 CNY in 2017.

4. Conclusions

As alluded to in the introduction, the satisfaction-based method has been commonly used in valuing ecosystem services, such as natural resources and environmental quality. In comparison with conventional nonmarket valuation methods, such as the travel cost and contingent valuation approaches, the satisfaction-based method has a number of advantages. These primarily relate to its survey design, which provides several desirable features, including simplified and less ambiguous questions in the survey, and the removal of the need to describe the public good to be valued, thus eliminating the possibility of strategic biases. Furthermore, the economic value derived using the marginal rate of substitution (MRS) between the numeraire variable of income/cost expenditure and the public good being valued based on a compensated demand function is straightforward and much easier to implement compared to the value derived using the CV method. For these reasons, among others, in this study we aimed to undertake novel research to propose the tourism satisfaction (TS) method for the valuation of important characteristics of the national forest parks in China.

The analytical results of our empirical analysis were based on a large sample dataset gathered from 22 NFPs and 4531 survey participants in China. The analysis leads us to the following conclusions: (i) The quality of the park recreational attributes has a significant effect on tourist satisfaction. Specifically, tourists attached the greatest importance to the rate of vegetation coverage, environment condition, and facilities offered at the park sites. Furthermore, the analysis also encompassed several park management factors, which showed tourists are sensitive to the status of the park's congestion because it has a significantly negative effect on tourist satisfaction. Thus, tourists exhibit a high level of willingness to pay to suppress the level of park congestion. (ii) There is a divergence in the preferences of local and nonlocal tourists regarding both the NFP attributes and their willingness to pay. For instance, nonlocal tourists pay more attention to the vegetation coverage and environmental quality than the local tourists; thus, the two attributes make a considerable contribution to the level of tourist satisfaction. In comparison, the local tourists attach more importance to the traffic conditions. Specifically, the nonlocal tourists are willing to pay as much as CNY 556.03 (USD 82.35) to reduce the level of congestion at the sites, whereas the local tourists' WTP for the same attribute is only CNY 117.47 (USD 17.39). (iii) The quality enhancement of the recreation attributes, and tourists' level of satisfaction, provide significant benefits to their travel experience.

The monetized values of the recreational attributes of NFPs can not only help policy makers understand the total economic value that can potentially accrue to the recreational services of the forest ecosystem, but they can also be used to derive some insightful implications for future NFP management. First, it appears that rubbish reduction and congestion control at the park sites are the most important tasks for the park management, thus suggesting that, in future resource allocations, including of both finance and labor, these two items should be targeted first. Second, the quantity of rubbish and tourist numbers at the NFP sites should be considered to be important indicators in ranking the NFPs, instead of placing too much emphasis on the parks' land sizes, which have been historically used as a criterion for ranking NFPs. Finally, the notable difference in the WTPs of the local and nonlocal tourists may constitute economic rationality for implementing differential park admission charges. Thus, the park management could impose a relatively higher price on outside tourists compared to local tourists. Given the digital technology that currently prevails in China, such as instant payments using mobile phones, the park authority would face few technical barriers or challenges to the implementation of such a discriminatory price policy. However, it should be noted that the revealed WTP of the nonlocal tourists should not be interpreted as the optimal admission fee to be charged by the NFPs. Rather, it should only be deemed to be a meaningful reference for the level of the fee.

Supplementary Materials: The following are available online at https://www.mdpi.com/article/10.3390/f12121688/s1, Table S1: The selected National Forest Park and its basic information, Table S2: The descriptive statistics result of individual characteristic.

Author Contributions: Data curation, N.K.; formal analysis, N.K. and Y.Y.; methodology, N.K.; software, N.K.; Writing—Original Draft, N.K.; Writing—Reviewing and Editing, N.K. and E.W.; Funding acquisition, N.K. and E.W.; Project administration, E.W. and Y.Y.; Supervision, E.W. and Z.D.; Data curation, Z.D. All authors have read and agreed to the published version of the manuscript.

Funding: This research was funded by the Humanities and Social Science Foundation of Ministry of education, grant number 21YJCZH057; the China's National Social Science Foundation, grant number 71640035; and Humanities and Social Sciences Project of Liaoning Provincial Department of Education, grant number JW202002.

Data Availability Statement: The data presented in this study are available on request from the corresponding author.

Conflicts of Interest: The authors declare no conflict of interest.

References

1. TEEB. *The Economics of Ecosystems and Biodiversity: Ecological and Economic Foundations*; Earthscan: London, UK; Washington, DC, USA, 2010.
2. Cardoso, P.; Barton, P.S.; Birkhofer, K.; Chichorro, F.; Deacon, C.; Fartmann, T.; Fukushima, C.S.; Gaigher, R.; Habel, J.C.; Hallmann, C.A.; et al. Scientists' warning to humanity on insect extinctions. *Biol. Conserv.* **2020**, *242*, 12. [CrossRef]
3. Del Saz-Salazar, S.; Navarrete-Tudela, A.; Alcalá-Mellado, J.R.; del Saz-Salazar, D.C. On the use of life satisfaction data for valuing cultural goods: A first attempt and a comparison with the contingent valuation method. *J. Happiness Stud.* **2019**, *20*, 119–140. [CrossRef]
4. Kant, S.; Vertinsky, I.; Zheng, B. Valuation of first nations peoples' social, cultural, and land use activities using life satisfaction approach. *For. Policy Econ.* **2016**, *72*, 46–55. [CrossRef]
5. Frey, B.S.; Luechinger, S.; Stutzer, A. The life satisfaction approach to environmental valuation. *Annu. Rev. Resour. Econ.* **2010**, *2*, 139–160. [CrossRef]
6. Levinson, A. Valuing public goods using happiness data: The case of air quality. *J. Public Econ.* **2012**, *96*, 869–880. [CrossRef]

7. Liu, R.; Wang, W. From revealed preference, stated preference to happiness data: The research method review OF Public goods valuation. *Econ. Rev.* **2014**, *2*, 150–160.
8. Welsch, H. Preferences over prosperity and pollution: Environmental valuation based on happiness surveys. *Kyklos* **2002**, *55*, 473–494. [CrossRef]
9. Ambrey, C.L.; Fleming, C.M.; Chan, A.Y.-C. Estimating the cost of air pollution in South East Queensland: An application of the life satisfaction non-market valuation approach. *Ecol. Econ.* **2014**, *97*, 172–181. [CrossRef]
10. Ferreira, S.; Moro, M.; Clinch, P.J. Valuing the environment using the life-satisfaction approach. *Plan. Environ. Policy Res. Ser. Work. Pap.* **2006**. [CrossRef]
11. Ambrey, C.L.; Fleming, C.M. Valuing scenic amenity using life satisfaction data. *Ecol. Econ.* **2011**, *72*, 106–115. [CrossRef]
12. Tsurumi, T.; Shunsuke, M. Environmental value of green spaces in Japan: An application of the life satisfaction approach. *Ecol. Econ.* **2015**, *120*, 1–12. [CrossRef]
13. Rehdanz, K. *Species Diversity and Human Well-Being: A Spatial Econometric Approach*; Hamburg University Centre for Marine and Atmospheric Science: Hamburg, Germany, 2007.
14. Krekel, C.; Zerrahn, A. Does the presence of wind turbines have negative externalities for people in their surroundings? Evidence from well-being data. *J. Environ. Econ. Manag.* **2017**, *82*, 221–238. [CrossRef]
15. Chen, C.F.; Fu, S.C. Experience quality, perceived value, satisfaction and behavioral intentions for heritage tourists. *Tour. Manag.* **2010**, *31*, 29–35. [CrossRef]
16. Jin, N.; Sangmook, L.; Hyuckgi, L. The effect of experience quality on perceived value, satisfaction, image and behavioral intention of water park patrons: New versus repeat visitors. *Int. J. Tour. Res.* **2015**, *17*, 82–95. [CrossRef]
17. Oh, H.; Flore, A.M.; Miyoung, J. Measuring experience economy concepts: Tourism applications. *J. Travel Res.* **2007**, *46*, 119–132. [CrossRef]
18. Sahin, I.; Ozlem, G.F. Do experiential destination attributes create emotional arousal and memory? A comparative research approach. *J. Hosp. Mark. Manag.* **2020**, *29*, 956–986.
19. Barbara, M. The service profit chain: How leading companies link profit and growth to loyalty, satisfaction, and value. *Int. J. Serv. Ind. Manag.* **1997**, *9*, 312–313.
20. Williams, D.R.; Vaske, J.J.; Kruger, L.E.; Jakes, P.J. The measurement of place attachment: Validity and generalizability of a psychometric approach. *For. Sci.* **2003**, *49*, 830–840.
21. Freeman, A.M. *The Measurement of Environmental and Resource Values: Theory and Methods*; Resources for the Future Press: Washington, DC, USA, 1993.
22. Welsch, H. Implications of happiness research for environmental economics. *Ecol. Econ.* **2009**, *68*, 2735–2742. [CrossRef]
23. Divisekera, S. Economics of tourist's consumption behaviour: Some evidence from Australia. *Tour. Manag.* **2010**, *31*, 629–636. [CrossRef]
24. Tisdell, C. Valuation of tourism's natural resources. In *International Handbook on the Economics of Tourism*; Edward Elgar Publishing: Cheltenham, UK, 2006.
25. Frey, B.; Stutzer, A. What can economists learn from happiness research. *J. Econ. Lit.* **2002**, *40*, 402–435. [CrossRef]
26. Reisinger, Y.; Turner, L.W. The determination of shopping satisfaction of Japanese tourists visiting Hawaii and the Gold Coast compared. *J. Travel Res.* **2002**, *41*, 167–176. [CrossRef]
27. Welsch, H.; Kühling, J. Using happiness data for environmental valuation: Issues and applications. *J. Econ. Surv.* **2009**, *23*, 385–406. [CrossRef]
28. Giergiczny, M.; Czajkowski, M.; Żylicz, T.; Angelstam, P. Choice experiment assessment of public preferences for forest structural attributes. *Ecol. Econ.* **2015**, *119*, 8–23. [CrossRef]
29. Hendee, J. *Wilderness Management*, 3rd ed.; Fulcrum Press: Golden, CO, USA, 2002.
30. Wang, E.D.; Wei, J.H.; Lu, H.Y. Valuing natural and non-natural attributes for a national forest park using a choice experiment method. *Tour. Econ.* **2014**, *20*, 1199–1213. [CrossRef]
31. Christie, M.; Hanley, N.; Hynes, S. Valuing enhancements to forest recreation using choice experiment and contingent behaviour methods. *J. For. Econ.* **2007**, *13*, 75–102. [CrossRef]
32. Sælen, H.; Ericson, T. The recreational value of different winter conditions in Oslo forests: A choice experiment. *J. Environ. Manag.* **2013**, *131*, 426–434. [CrossRef]
33. Chaminuka, P.; Groeneveld, R.A.; Selomane, A.O.; Van Ierland, E.C. Tourist preferences for ecotourism in rural communities adjacent to Kruger National Park: A choice experiment approach. *Tour. Manag.* **2012**, *33*, 138–172. [CrossRef]
34. Jarvis, D.; Stoeckl, N.; Liu, H.-B. The impact of economic, social and environmental factors on trip satisfaction and the likelihood of visitors returning. *Tour. Manag.* **2016**, *52*, 1–18. [CrossRef]
35. Hill, R.J.; Melser, D. Hedonic imputation and the price index problem: An application to housing. *Econ. Inq.* **2008**, *46*, 593–609. [CrossRef]
36. Ferrericarbonell, A.; Van Praag, B.M.S. *Happiness quantified: A satisfaction calculus approach*; Oxford University Press: Oxford, UK, 2004.
37. Ferreira, S.; Moro, M. On the use of subjective well-being data for environmental valuation. *Environ. Resour. Econ.* **2010**, *46*, 249–273. [CrossRef]

38. Penn, J.; Hu, W.Y.; Cox, L.J.; Kozloff, L. Values of recreational beach quality in Oahu, Hawaii. *Mar. Resour. Econ.* **2015**, *31*, 47. [CrossRef]
39. Krinsky, I.; Robb, A.L. On approximating the statistical properties of elasticities. *Rev. Econ. Stat.* **1986**, *68*, 715–719. [CrossRef]
40. Lindberg, K.; Veisten, K. Local and non-local preferences for nature tourism facility development. *Tour. Manag. Perspect.* **2012**, *4*, 215–222. [CrossRef]

forests

MDPI

Article

Valuation of Forest Ecosystem Services in Taiwan

Jiunn-Cheng Lin [1], Chyi-Rong Chiou [2], Wei-Hsun Chan [3] and Meng-Shan Wu [3,*]

1. Chief Secretary Office, Taiwan Forestry Research Institute, No. 53 Nanhai Rd., Taipei 10066, Taiwan; ljc@tfri.gov.tw
2. Department of Forestry and Resource Conservation, National Taiwan University, No. 1, Sec. 4, Roosevelt Rd., Taipei 10617, Taiwan; esclove@ntu.edu.tw
3. Forestry Economics Division, Taiwan Forestry Research Institute, No. 53 Nanhai Rd., Taipei 10066, Taiwan; frog@tfri.gov.tw

* Correspondence: wumengshan@tfri.gov.tw; Tel.: +886-2-2303-9978 (ext. 1325)

Abstract: Forest is the largest ecosystem in the land area of Taiwan. In the past, most of the studies on the evaluation of forest ecosystem services were regional, and therefore lacked national assessment. This study uses a market value method and a benefit transfer method to assess the value of the forest ecosystem services in Taiwan, and expounds the link between ecosystem services and the effectiveness of forestry management and conservation. Preliminarily, it is estimated that the total value of forest ecosystem services in 2016 was approximately NT $749,278 million (equal to approximately 47.6 billion U.S. dollars, PPP-corrected), accounting for 4.28% of the GDP in 2016. The quotation of unit price data has a huge impact on the final assessment results of forest ecosystem service value, and therefore it is necessary use it appropriately.

Keywords: forest ecosystems; ecosystem services; economic valuation

Citation: Lin, J.-C.; Chiou, C.-R.; Chan, W.-H.; Wu, M.-S. Valuation of Forest Ecosystem Services in Taiwan. *Forests* **2021**, *12*, 1694. https://doi.org/10.3390/f12121694

Academic Editors: Elisabetta Salvatori and Giacomo Pallante

Received: 16 November 2021
Accepted: 2 December 2021
Published: 3 December 2021

Publisher's Note: MDPI stays neutral with regard to jurisdictional claims in published maps and institutional affiliations.

1. Introduction

The concept of ecosystem services (ESs) has gained recognition and been put into use since the late 1970s [1], and ES classification methods have also been proposed gradually [2–4]. The "Millennium Ecosystem Assessment (MEA)" of 2005, which assessed the effects of ecosystem changes on human well-being, laid a scientific foundation for the actions required to strengthen ecosystems' contributions to human well-being, without damaging their long-term productivity. The 2010 Economics of Ecosystems and Biodiversity (TEEB) analysis was based on the relevant research field of the past few decades, and proposed a method that can help decision makers recognize, demonstrate and—under appropriate circumstances—capture the value of ecosystems and biodiversity. The Common International Classification of Ecosystem Services (CICES) provides more formal and systematic definitions of various types of ecosystem services, presenting the pathways of ecosystem services from ecological structure and processes to human well-being, and helping people easily identify the content covered by different service categories. The classification method can be converted between MA, TEEB and other frameworks for the purpose of natural capital accounting. The United Nations has now developed the System of Environmental–Economic Accounting—ecosystem accounting [5]—which aims to link information about natural assets with economic and other human activities. Wealth Accounting and the Valuation of Ecosystem Services (WAVES) aims to promote sustainable development by ensuring that natural resources are mainstreamed in development planning and national economic accounts [6].

In order to increase certain specific ESs in an increasingly resource-constrained world, other ecosystem services have been degraded as result [2,7]. One of the main reasons for the decrease in ecosystem services is that their true value is not considered in economic decision-making [2]. Most decisions are based on market prices, but for many ecosystem services, there is no market, and decision makers have no clear signal on the value of the service [8]. A better understanding of the economic value generated by ESs can facilitate the adoption of efficient policies and measures to preserve and enhance them [9]. The traditional evaluation methods of ecological services and economic services include revealed and stated preferences [10]. In order to carry out research on ecosystem services, the benefit transfer method, which finds empirical economic valuation data under various backgrounds and spatial scales, and adjusts them according to the differences in ecological and economic backgrounds, is also a common evaluation method [11,12]. The natural capital associated with ecosystem services is inherently uncertain, and its decision-making involves many stakeholders in the economic evaluation of value, which makes imperfect information a key challenge for economic evaluation [13]. Current ecosystem services studies focus on developing quantifiable, replicable and credible methods that can explicitly consider biophysical, socio-cultural and monetary value fields, so that decision makers can incorporate them into the decision-making process, and link the biophysical mechanisms provided by ESs with the socio-economic impact of their use. [14–16].

The forest ecosystem not only provides basic living materials such as wood, fuel and recreational services for human beings, it also provides fundamental functions of life, such as biological gene banks, animal and plant habitats, climate regulation, air purification, and carbon sequestration, etc. [17,18]. Most of the existing research work focuses on provisioning, regulating and cultural services that have a relatively well-developed methodology. Other services, such as pollination, genetic resources and gene pool protection, regulation of pests and human disease, the aesthetic values of forests, waste treatment, environmental purification, and disease regulation, have received less attention in the scientific community, due to lack of data, challenges in estimating their value, and lack of well-designed methods, among other reasons [19]. China [20], Japan [21] and South Korea [22], adjacent to Taiwan, have successively completed national forest ecosystem service value assessments, all of which have achieved considerable results that have been applied to forestry policy communication and policy benefit evaluation. Forests are the largest ecosystem in the land area of Taiwan. In the past, most studies on the evaluation of the service value of forest ecosystems were regional, or focused on services such as conservation, soil and water conservation, and carbon sequestration [23–25], and lacked national forest service value. Although, awareness of the importance of ecosystem services to human beings is growing, and people are beginning to understand that ecosystem services need to be taken seriously. It is not easy to provide reliable and well accepted results when performing an economic assessment of ecosystem services. The main purpose of this study is to confirm the evaluation items and evaluation methods of forest ecosystem services in Taiwan, and to estimate their multi-oriented service value. Thus, we can provide decision-makers with opportunities to re-examine environmental resources and their values, so that ecosystem services can be more easily incorporated into public decision-making processes.

2. Materials and Methods

2.1. Research Area

Taiwan is a long and narrow island stretching from north to south, with a total land area of 36,000 square kilometers, located in East Asia on the southwestern edge of the Pacific Ocean (Figure 1a). The average annual precipitation of the island is about 2221 mm on flatland, and about 3858 mm in mountainous areas, with an average annual temperature of about 20–25 °C. Taiwan has many high mountains, mainly distributed in the east of the island, with a vertical height difference of nearly 4000 m. The ecological environment is diverse, with rich biodiversity and a high proportion of endemic species and subspecies. There are more than 5000 species of vascular plants in the wilderness of Taiwan, about a quarter of which are endemic [26]. The total forest land of Taiwan is approximately 1,868,636 hectares, which includes 1,533,691 hectares of state-owned forest (81.07% natural forest and 18.93% planted forest), and 334,944.63 hectares of public and private forest (Figure 1b), which are mostly located in the shallow hills connected with the flatlands. The state-owned forest is divided into 36 national working circles (Figure 1c). In addition to forest land, the total forest land area also includes 1,695,469 hectares of woodland. The areas of Taiwan protected for the purpose of nature conservation consist of 6 categories, including national parks, national natural parks, nature reserves, natural reserves, wildlife reserves, and important wildlife habitats, totaling 95 protected areas. The total land area of these categories is 694,503 hectares (Figure 1d), accounting for approximately 19.19% of the land area of Taiwan [27].

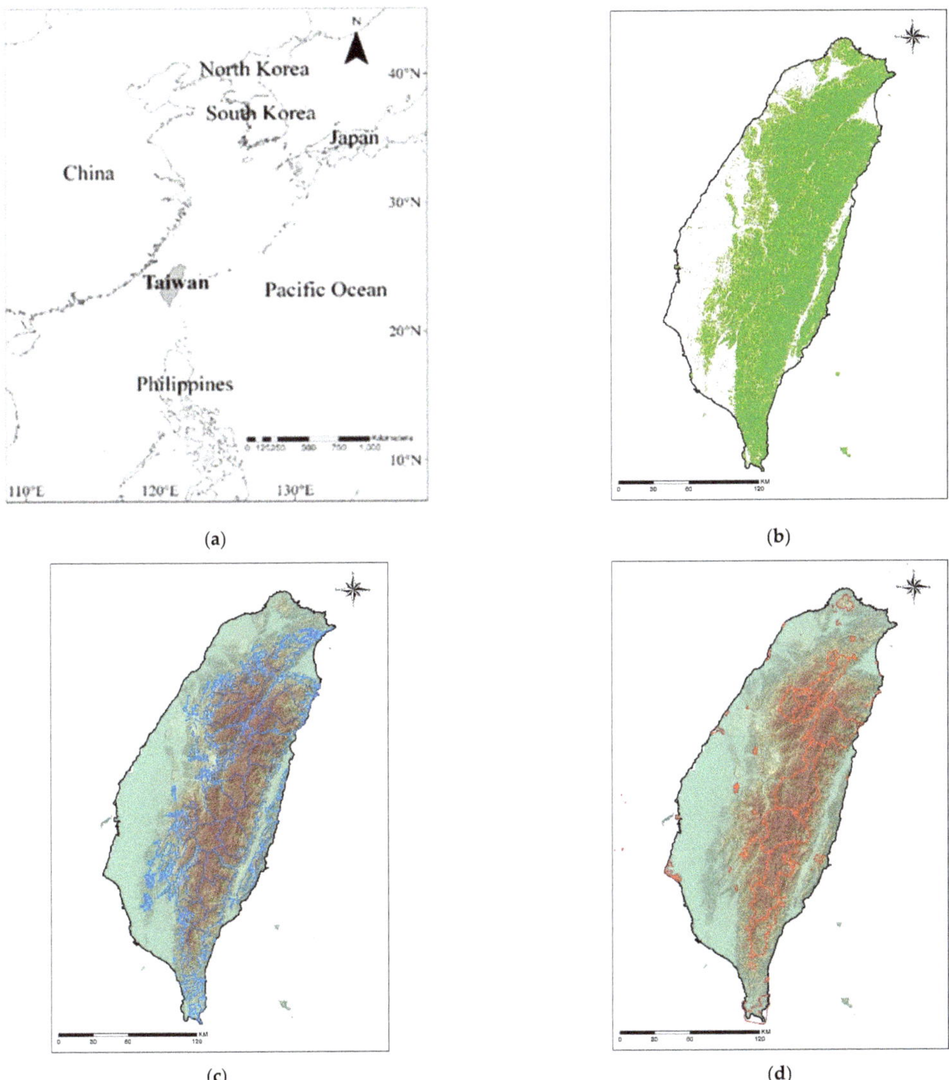

Figure 1. The geographical location and forest resources of Taiwan. (**a**) Geographical location of Taiwan. (**b**) Forest coverage of Taiwan. (**c**) The national forest working circles in Taiwan (area marked with blue lines). (**d**) Protected areas in Taiwan (area marked with red lines).

2.2. Research Methods

In order to conduct a nationwide ecosystem service value assessment, this study selected China, Japan, South Korea, etc., as countries that have completed national forest ecosystem service value studies [20–22]. Then, we analyzed their evaluation items and methods as a reference for the choice of ecosystem service evaluation items and method in Taiwan. More than two countries listed water resources conservation, water quality purification, soil and sand loss prevention and control, carbon sequestration, air purification, biodiversity, forest recreation, etc. as assessment items. The evaluation methods were mainly based on market price-based approaches, cost-based approaches, and the benefit transfer method. This study refers to [20–22] to summarize common assessment items, and considered factors such as the data source and parameters of the assessment items, and the ease of obtaining information. In addition to water purification, the other six ecosystem services were used as assessment items in Taiwan.

The ecosystem service valuation considered factors such as the idiosyncratic roles of different services, the difficulties of obtaining information, the cost of survey and the accuracy of the evaluation. The evaluation methods varied: direct market valuation was conducted for general provisioning services; regulating services were mainly evaluated by avoiding cost, replacement cost or contingent valuation methods; cultural services were evaluated using travel cost methods (such as entertainment, tourism, or scientific value), hedonic methods (such as aesthetic value), or contingent valuation (such as existence value) [28]. In the absence of detailed ecological and economic data, the benefit transfer method is a common approach used to transfer the existing value through a similar ecosystem (research location), to estimate the service value of the ecosystem (designated location by policy) [29,30].

This study mainly used market valuation methods, cost-based methods and benefit transfer methods to assess the value of forest ecosystem services. In terms of quantifying the intensity of ecosystem service activities (measurement), current state or data available publicly were selected as research data. In terms of quantifying the intensity of ecosystem service activities, we selected current state or publicly available data as research data. Two expert meetings were also held, and scholars with expertise in soil and water conservation, carbon sequestration, air purification, forest recreation and biodiversity were consulted to discuss and confirm the data sources, parameters and other information of the forest ecosystem service assessment project. As for the evaluated unit price, experts in related fields were also invited to conduct two expert meetings to determine the unit price. About 4~6 experts participated in each meeting. The evaluation methods, indicators and unit prices of different forest ecosystem services are shown in the Table 1.

Table 1. Ecosystem service valuation method, indicator and unit price used in this study.

Ecosystem Service	Valuation Method	Indicator	Unit Price
Forest water conservation	Market price-based approaches	Water conservation quantity(m^3)	Shadow price of industrial water
Air purification	Cost-based approaches	Removal of pollutants (kg)	Pollution control cost
Biodiversity	Benefit transfer method	Protected area (ha)	Benefits of protected areas
Soil loss prevention	Cost-based approaches	Soil loss (m^3)	Costs of river dredging and reservoir dredging
Forest carbon sequestration	Market price-based approaches	Carbon storage (t)	Carbon price
Forest recreation/environmental education	Benefit transfer method	Forest recreation visits /Person-time of forest environmental education	Forest recreation price /Forest environmental education price

2.3. Ecosystem Services Valuation

2.3.1. Water Conservation

As rivers in Taiwan are short and rapid, reservoirs must be used to retain water resources, otherwise precipitation will flow into the sea in a short time. Forests can delay the loss of water resources, reduce the loss of unused water resources, and ensure a more efficient use of water resources. Water conservation means that forests can intercept, absorb and store precipitation, and convert surface water into surface runoff or groundwater. Water conservation generally uses the water-balance method, base-flow estimation method, and retreat-curve method to estimate the water conservation capacity of each watershed area. However, because the flow, evapotranspiration and other data at domestic watershed areas are incomplete, the water-balance method is used to estimate the amount of water conservation. The hydrological balance method regards the forest ecosystem as a closed-loop cycle, thereby directly considering the input and output of water in the soil, which is widely used in the calculation of many water resources. The discrepancy between precipitation and consumption, such as forest evapotranspiration, is the forest water conservation (water-retention capacity), which is then multiplied by the forest land area to get the total forest water conservation quantity, as seen in Equation (1):

$$w = \Sigma(R/1000) \times \theta \times A \quad (1)$$

where w is water conservation quantity (metric t), R is average annual precipitation (mm/year), θ is empirical formula of water conservation ratio (%), and A is working circle (m^2) (there are 36 national forest working circles in Taiwan). Water conservation ratio in Taiwan during dry and wet periods is about 19–30% of rainfall, so an average of 25% is used as the conservation ratio.

Regarding the unit price of water conservation valuation, the case study of water price assessment in Japan [31] uses reservoir maintenance costs (including construction and maintenance costs) as replacement costs. China [32] regards the cost of reservoir unit volume as the replacement cost, and the cost of different reservoirs varies greatly, about 4.25–30.20 NT dollars/m^3. In Taiwan, [33] estimates the value of forest water conservation based on the tap water price of 6.6 NT$/$m^3$. Moreover, [34,35] estimate the value of water conservation based on the shadow price of water used by the manufacturing industry in NT $21.56 and NT $15.36 respectively. This study originally intended to use water price as the unit price, but after discussion at the expert meeting, it was believed that the current water price could not reflect the cost of water use. Therefore, the shadow price of industrial water NT $40.81/$m^3$ investigated by [36] was used as the unit price (P_w). The value of water conservation is calculated by Equation (2):

$$V_w = W \times P_w \quad (2)$$

where V_w is water conservation value (NT), W is water conservation quantity (metric t), and Pw is price per metric t of water conservation (NT/m^3).

2.3.2. Air Purification

Air purification refers to the ability of forest trees to trap pollutants on the surface of the leaves through the canopy when it is not raining, which is the function of adsorbing and purifying air pollutants [37,38]. In Taiwan, the net throughfall method is mainly used to estimate the amount of air pollution removed directly from the atmosphere by forest trees. The net throughfall method uses the difference between the input of ion components in the rain outside the forest and the throughfall of the forest canopy to estimate air pollutants intercepted by forest stand. Since the forest canopy can effectively intercept the dry deposition, which will be washed by rainwater to the ground surface, the difference in the contents of the elements between throughfall and the precipitation can be estimated [39]. Another study cited by this study investigates how forests in Taiwan contribute to the purification of air pollutants, monitoring the removal amount of particulate pollutants,

SO_2, and NOx, in flatland afforestation [40–42]. The monitoring period lasted more than 1 year, so the factors that could affect the removal amount such as season, fallen leaves and non-fallen leaves were able to be considered. Consequently, the study used the average removed amount by different tree species as the value of air pollutants adsorbed by the forest stand. The average annual removal of particulate pollutants, SO_2, and NOx per hectare was 47.79, 7.17, and 6.52 kg respectively.

Using the cost for pollution prevention as a replacement cost is a benefit of the role of the forest as an air pollution filter. Thereby, the cost of using artificial technology to replace the cost generated by ecosystem services can be estimated. Due to the idiosyncratic variation in pollution sources, the pollution prevention methods and costs used varied accordingly [43]. The unit reduction cost of different pollutants was based on the latest available measures provided by the Environmental Protection Administration of the Executive Yuan [44], and was adjusted annually with the deflator of the pollution remediation industry [45]. The benefits of air pollutant removal of the forest in 2016 were estimated, using per unit leaf area particulate adhesion capacity × forest coverage area (with trees) × pollution prevention cost, as seen in Equation (3):

$$V_{\text{Air purification}} = (Q_{k1} + Q_{k2} + Q_{k3}) \times A \times \left(P_{kf} + P_{km}\right)/2 \tag{3}$$

where $V_{\text{Air purification}}$ is the value of forest stand capacity to adsorb various pollutant. Q_{k1}, Q_{k2} and Q_{k3} are the amount of particulate pollutants, sulfur dioxide and nitrogen oxides removed per unit area of the forest stand (kg/ha/year). A is forest coverage area (ha), and P_{kf} and P_{km} are the cost per unit of prevention and control of particulate pollutants, sulfur dioxide, and nitrogen oxides (NT$/kg), when the pollution source is fixed or mobile, respectively. In the absence of information such as the relationship between pollution sources and forest land, this is a more reasonable way to estimate the range of air purification benefits based on the prevention and control costs of fixed and mobile pollution sources. In this study, the cost of fixed sources and mobile sources was averaged to calculate the air purification benefits, which was an expedient measure for the simple presentation of the evaluation results.

2.3.3. Biodiversity

Biodiversity refers to the function of the forest as a habitat conservation site to maintain and support species in range [3]. In view of the considerable differences in the assessments of biodiversity values in many countries, and of the lack of representative data on biodiversity conservation, the benefit transfer method is used to select representative indicators for evaluation. Presently, the development of biodiversity indicators is receiving more attention in the world. Biodiversity indicators can reflect whether an ecosystem is healthy and resilient, and monitor the status and trends of biodiversity [46]. For example, the "forest coverage area" is related to ecological, social, economic, and cultural impacts, and can be used as an indicator of land use and development pressure. Taiwan has established the "Taiwan Biodiversity Observation Network (TaiBON) [47]" with a stringent attitude, and has exchanged discussions with a diverse range of experts. Meanwhile, we have been referring to the developmental experience and framework of international biodiversity indicators to compile a set of biodiversity indicators suitable for Taiwan. Considering the comprehensiveness of current data collection and the frequency of updates, as well as existing research on relevant benefit evaluation, this study prioritizes the selection of "protected areas" as an indicator of biodiversity, which corresponds with the Sustainable Development Goals (SDGs) 15.4 of the United Nations and Aichi Target 11. The conservation price per unit area, referring to the conservation benefit of domestic nature reserves surveyed by [48], is about NT $79,075 per hectare, as seen in Equation (4).

$$V_B = I_B \times P_B \tag{4}$$

where V_B is biodiversity value (NT), I_B is protected area (ha), and P_B is the benefit of protected areas per unit area (NT).

2.3.4. Soil Erosion Prevention

The prevention and control of soil erosion means that covering the land surface with forest vegetation, and humus or fallen leaves produced by trees can reduce the direct erosion of the soil by rainfall, thereby reducing the rills and gullies in the ground eroded by surface water flow. In addition, tree roots contribute to a well-granulated soil with more frictional resistance, increasing the permeability of the soil. The migration and sediment deposit of forest soil, due to natural factors such as rainfall or river flow paths, and topographical factors such as slope gradient, is called soil erosion. Taiwan is surrounded by the sea on all sides, and the terrain is towering and steep, with plenty of rainfall. After rainfall, most water flows to the sea; therefore, the accumulated soil erosion over the years is considerable. This study uses the modified universal soil erosion formula, Revised Universal Soil Loss Equation (RUSLE) [49], to conduct the evaluation. First, we calculated the amount of soil loss for both scenarios, wherein there was woodland and bare land. Second, we subtracted the two kinds of losses, to obtain the reduced soil loss due to the existence of forest land, and then multiplied the soil loss by the SDR (sediment delivery ratio) in each watershed, to obtain the actual soil production of each watershed area. This is the amount of soil erosion that the forest can reduce, as seen in Equation (5):

$$Q_s = \sum_{j=1}^{36} \left(A_{m(F)} - A_{m(NF)} \right) \times \text{SDR} \times A_j \tag{5}$$

where $A_m = R_m \times K_m \times L \times S \times C \times CPF$, Q_s is average annual soil loss per hectare (metric t/ha-year), $A_{m(F)}$ is average annual soil loss per hectare from forest land (metric t/ha-year), $A_{m(NF)}$ is the average annual soil loss per hectare from bare woodland (metric t/ha-year), R_m is the erosion index of average annual rainfall (106 Joule-mm/ha-hour-year), K_m is the soil erosion index (metric t-ha-hour-year/106 joule-mm-ha-year), L is slope length, S is slope gradient, C is coverage and management, CPF is the water and soil conservation factor, SDR is sediment delivery ratio (%), which refers to the ratio of the yield of mud and sand in the watershed area to the amount of soil erosion, and Aj are the 36 national working circles (ha).

The avoided cost method was used to calculate the price of soil erosion prevention and control. By considering the problems of sedimentation and sand transport in rivers, the costs of river and reservoir dredging are used instead to estimate the total amount of soil erosion with the general soil erosion formula. The amount of soil and sand left in the riverbed, and that flown into the reservoir, was calculated at 70% and 30%, respectively. Then, we collected the unit benefits of the rivers and reservoirs, and used it to estimate the value of soil and sand loss control, as seen by Equation (6):

$$V_s = Q_s \times P_s \tag{6}$$

where V_s is the value of soil erosion control (NT-year), Q_s is average annual soil erosion per hectare (metric t/ha-year), and P_s is unit price for forest soil erosion prevention (NT/metric t).

2.3.5. Forest Carbon Sequestration

The guidelines for National Greenhouse Gas Inventories promulgated by the Intergovernmental Panel on Climate Change (IPCC) of the United Nations specifically regulates the methodologies of carbon emissions/sequestration. Taiwan also calculates the total carbon removal of forests according to the IPCC method (Equation (7)) [50], including the carbon removal of forest land and the maintenance of forest land, the carbon emission of forest land and the maintenance of forest land, and the carbon removal caused by conversion of other land to forest land:

$$\triangle CO_2 = \triangle CO_{2\,G} + \triangle CO_{2\,L} + \triangle CO_{2\,G(\text{Increased woodland})} \tag{7}$$

here $\triangle CO_2$ is total carbon dioxide removal (Mt CO_2e), $\triangle CO_{2\,G}$ is carbon removal from forest land maintained by forest land, $\triangle CO_{2\,L}$ is carbon emissions of forest land maintained by forest land, and $\triangle CO_{2\,G(\text{Increased woodland})}$ is carbon removal from other land converted to forest land.

The estimation of forest carbon sequestration value is calculated by multiplying the total carbon removal amount from the forest by the carbon price per t, as seen in Equation (8):

$$V_c = \triangle CO_2 \times P_c \tag{8}$$

where V_c is forest carbon sequestration value (NT/Mt/year), $\triangle CO_2$ is the total carbon dioxide removal (Mt CO_2e), and P_c is the carbon price per t (NT/Mt). According to the discussion at the expert meeting, the social cost of carbon is an estimate of the monetary value of the global damage caused by man-made carbon dioxide emissions. The price of the forest carbon emissions trading market alone cannot reflect the cost of the entire society; therefore, the shadow price of the European Bank (2014) [51] for carbon reduction financing is used to estimate the value of forest carbon sequestration.

2.3.6. Forest Recreation (Including Environmental Education)

The service provided by the forest ecosystem with places for leisure and entertainment for humans is called forest recreation. Meanwhile, forests can also offer space for environmental education services. This study uses the benefit transfer method to evaluate the value of forest recreation (environmental education) by using forest recreation (environmental education) person-time × forest recreation (environmental education) price, as seen in Equation (9):

$$V_R = \Sigma N_R \times P_R + \Sigma N_E \times P_E \tag{9}$$

where V_R is the value (NT/year) of forest recreation (including environmental education), N_R is annual forest recreation visits (person/year), P_R is forest recreation price per person (NT/person), N_E is person-time of forest environmental education per year (person/year), and P_E is forest environmental education price per person (NT/person). The annual total number of tourists in each forest recreation area can be obtained from the statistical information of the Tourism Bureau. When determining the price of forest recreation, we referred to the survey data for domestic tourism from the Tourism Bureau of Taiwan [52], and considered an overnight stay or an extended stay. Hence, the cost for forest recreation per person is decided on the average cost per person per trip/the average number of days of stay. Regarding the number of participants in forest environmental education, the forestry statistical data of the Forestry Bureau are used as reference. The literature review shows that the Taiwanese willingness to pay (WTP) for environmental education and guided tours related to forests or ecological recreation is roughly in the range of 100–170 NT per person [53,54]. This study uses a minimum of 100 NT per person [52] on a conservative estimate for the forest environmental education pricing.

3. Results

Based on the evaluation methods of the six ecosystem services mentioned above, the forest ecosystem services value of Taiwan in 2016 was calculated as follows.

3.1. *Value of Forest Water Conservation*

Geographic information system (GIS) was used to overlay the rainfall data of the 36 state-owned forestry working circles from the Central Weather Bureau and the rainfall measurement stations of the Water Resources Agency, to calculate the average annual rainfall and annual water conservation in each working circle. It was estimated that the total water conserved in the forest land of Taiwan in 2016 was 13,876.14 million metric t.

This study set 40.18 NT/m^3 (medium-cost situation) as the price for water conservation. The value of water conservation in 2016 was calculated by multiplying the amount of water conservation and the unit price, which was NT $557,543 million.

3.2. Air Purification Value

This study refers to the study of the purification effect of forest trees in Taiwan on air pollutants: the average annual removal of particulate pollutants, SO_2 and NO_x per hectare is 47.79, 7.17, and 6.52 kg respectively [40–42].

Based on the latest unit reduction cost of each pollutant provided by the Environmental Protection Administration of the Executive Yuan, the pollution remediation industry deflator was used for adjustment annually, by utilizing the amount of adsorbed air pollution per unit area of forest × forest coverage area (with trees) × pollution prevention cost to estimate the benefits of forest removal of air pollutants in 2016. When using fixed source reduction costs as the unit price, the benefit was NT$1894 million. Alternatively, the benefit was NT$4684, using the mobile pollution sources as the unit price. Hence, the benefit of forest air purification was the average of the two, which was NT$3289 million (Figure 2).

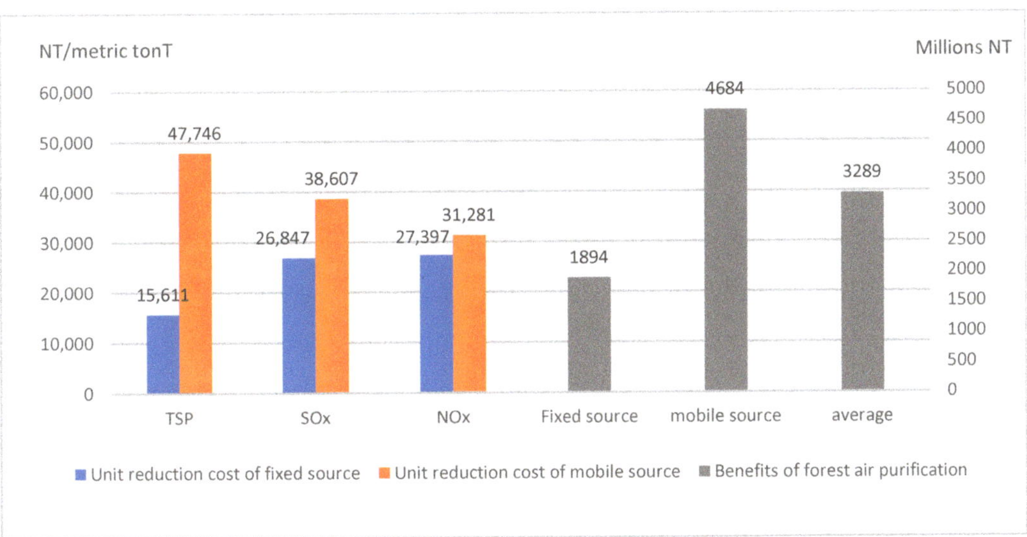

Figure 2. Unit reduction costs of different pollution sources and the benefits of forest air purification. Total suspended particles (TSP) (referring to particles suspended in the air) and suspended particles (PM_{10}) (referring to particles with a size of less than 10 μm) belong to the category of granular pollutants, so the unit control cost of TSP (total suspended particles) is used as that of particulate pollutants. Currently, the data available for per unit control cost is limited to SO_x, so the unit control cost of SO_2 is replaced by that of SO_x. Forest coverage area of woodland is 1,695,469 ha.

3.3. Biodiversity Value

The conservation areas in Taiwan are mainly based on the land protected areas currently announced, including national parks, national natural parks, nature reserves, natural reserves, wildlife reserves, and important wildlife habitats. There are 95 protected areas in 6 categories. In 2016, the protected area was 6945 km^2. According to [48], the conservation benefit of the domestic nature reserves was about NT $79,075 per hectare. This price was adjusted to the conservation benefit per unit of each year by applying the consumer price index, and then multiplied by the area of the protected area to obtain the biodiversity value (the conservation benefit of the protected area), which was NT $58,535 million.

3.4. Soil Erosion Prevention Value

The parameters in the analysis were calculated using grid data, among which the relevant L (slope length) and S (slope gradient) were converted using the 20 m-resolution topographic data of the whole station, published by the Ministry of the Interior. The R and K parameters were drawn into a gridded map with a resolution of 40 m, using data from the Water and Soil Conservation Handbook [55]. Finally, the C and P parameters were converted into 20 m grids, according to the land cover type. The results showed that the average amount of soil retained by the working circle in Taiwan was about 519 metric t/ha, and the total amount of soil erosion control was 719.73 million metric t.

The river silt reduction benefit was based on the direct benefit calculation of soil control by the reference manual for the overall survey and planning work of the watershed area [56], with an average about 112.5 NT/m^3. The efficiency of the reservoir silt reduction was based on the unit price of mechanical dredging, which averaged about 350 NT/m^3. The bulk density of soil per metric t was approximately 0.7 m^3, and the value of soil erosion prevention was NT $92,575 million.

3.5. Forest Carbon Sequestration Value

The 2018 Greenhouse Gases Inventory Report of Taiwan showed that the total carbon removal from forest in 2016 was 21,418 $\times$ 10^3 Mt CO_2e. The European Bank's shadow price for carbon reduction financing in 2014 was 35 Euros/Mt CO_2e. Reflecting this, the marginal cost of abatement will increase over time with the higher concentration of carbon dioxide in the atmosphere: the Bank of Europe sets the actual rate of increase in the shadow price every year as 2.0%. Therefore, it can be concluded that the value of forest carbon storage in 2016 was NT $27,879 million.

3.6. Value of Forest Recreation (Including Environmental Education)

The value of forest recreation includes forest recreation (A) and environmental education (B). Based on official statistics from the Tourism Bureau and the Forest Service Bureau, as well as the results of domestic research on willingness to pay for environmental education, the value of forest recreation (including environmental education) in 2016 was NT $9457 million (Table 2).

Table 2. Value of forest recreation (including environmental education).

Item	People/Year	Price	Benefit (NT Million)
Forest recreation (A)	6,133,885	1449 [1]	8888
Environmental education (B)	5,331,244	106.68 [2]	569
Forest recreation (including environmental education) (A + B)			9457

[1] The average cost per person per trip/the average number of days of stay [52].[2] Willingness to pay was based on the 2008 research data [54] as a base period of 100 for price adjustments.

3.7. Results of Forest Ecosystem Service Valuation of Taiwan

This study assessed the value of six different kinds of forest ecosystem services in Taiwan. The total value in 2016 was approximately NT$ 749,278 million (equal to approximately 47.6 billion U.S. dollars, PPP-corrected), accounting for about 4.28% of the GDP of 2016. Forest water conservation value was about NT $557,543 million, accounting for 74.41% of the annual total forest ecosystem service value ratio, which was the highest, followed by the prevention and control of soil erosion value at NT $92,575 million, accounting for 12.36% of the annual total forest ecosystem service value. The biodiversity value was about NT $58,535 million, accounting for 7.81% of the annual total forest ecosystem service value. In 2016, the total forest area of the country was 1,868,636 hectares. If the total value of forest ecosystem services in the year was divided by the total forest area of Taiwan, the value of the forest ecosystem services per hectare was NT $400,976.

4. Discussion

4.1. Research Limitations

There are still some issues that require attention in the acquisition and application of data sources in this study. First, the value of forest ecosystem services was derived by multiplying the value of each ecosystem service by the unit price, but the acquisition of relevant data in Taiwan was not complete. For example, the range of different forest ecosystem services are slightly different; some of the benefits are in the national forest working circle, but some cover forests throughout Taiwan. In terms of the benefits of forest air purification, the analysis of this study found that the distance from the pollution source affects forest air purification [57]. However, because there are very few studies regarding air purification of forests in Taiwan, the results of the purification of air pollutants by flatland afforestation survey were applied to estimate forestlands island-wide, which was a stopgap measure due to lack of information. As for biodiversity value, due to the limited quantitative indicators of biodiversity benefits, in order to facilitate the Forestry Bureau to continuously update and evaluate their information, currently only one research result is cited. The estimation was carried out by the benefit transfer method, and the scope of estimation was limited to the conservation benefit of the natural conservation areas. With respect to water conservation, bogged down by currently incomplete data regarding rainfall and surface runoff, and the upstream of each watershed area, the use of relevant research to estimate the water conservation ratio in the watershed areas of the mountains of Taiwan still lacks accuracy. Forest recreation is currently based on the average daily travel expenses per person in the Tourism Bureau as the price unit, but this value does not distinguish between consumption in forest and non-forest areas. Further evaluation methods, discussion and correction should be carried out in the future.

In terms of unit price quotation, due to the lack of updated research data, the results of this study are outdated [44,48]. In addition, due to the lack of data on air pollution sources, the average value of the reduction costs of mobile pollution sources and fixed pollution sources was used to estimate the value of air pollution. To reasonably present the value of ecosystem services, the task of developing and updating data integrity requires our continuous effort to overcome challenges. Furthermore, the result of this study was on a larger scale, so its applicability is a concern. To draw upon it in a case or regional study, more complete and detailed information is required, as the supply and demand contribution of each case is different [10]. It is suggested that more regional research can be done on case studies.

4.2. Price Difference

Conducting forest ecosystem services evaluation requires a great deal of judgement and objectivity using unit price data; without this, there could be a significant impact on the final evaluation result of the forest ecosystem service value. For example, in the estimation of the value of water conservation, the shadow price of industrial water use is four times the current water price, and the difference between the two estimated values will be four times as much. In China [58] and Japan [21], the costs of constructing, maintaining and operating reservoirs for public water supply and depreciation are mainly used as replacement costs to estimate the preservation value of water bodies. South Korea [22] also considers the opportunity cost of land use for dam construction, such as the opportunity cost of planting rice and forests. Relatively speaking, the current price for calculating the value of water conservation in Taiwan is higher than these countries. The price of forest carbon storage, citing the shadow price of the European Bank of Carbon Finance, also differs about 14 times from the price in the international voluntary carbon trading market [59]. The shadow price reflects the marginal contributions of assets to well-being [60]. In the national accounts, the exchange value is easier to observe and estimate than the shadow price, while ensuring the integrity and consistency of the account [61]. Both should be selected carefully, according to the purpose.

4.3. Development for Biodiversity Assessment Methodology

Biodiversity and ecosystems are conceptually linked in three ways: (1) biodiversity is the support and foundation of ecosystems; (2) conflicts lie between biodiversity and ecosystems; (3) biodiversity means ecosystem services (or part of it) [62]. The corresponding service items vary in different ways, which also affects the evaluation of the biodiversity value. This study used only the natural reserve areas to estimate the value of biodiversity conservation, which simplified the relationship between biodiversity and ecosystem services.

Every country differs greatly in their biodiversity value assessment indicators. China uses the concept of conserving and maintaining the biodiversity of the forest ecosystem for hunting and gathering as its value calculation basis. The Shannon–Wiener index is used to classify the different biodiversity levels of the forest: the higher the biodiversity, the higher the value. The biodiversity value is calculated based on the wild animal market price and the forest product replacement cost to estimate the integrated price [63,64]. Japan estimates the bird population size in the forest based on the density of the forest that forms bird habitat. The area of each forest forms a biodiversity indicator, and then its value is calculated based on the price of artificial feeding bait, whereas South Korea's biodiversity value is based on the sum of the value of wild bird conservation, and that of hunting activities [31,65].

The current "TAIBON" indicators [47] are complete and diverse, but many biodiversity indicators lack time continuity. Moreover, although some monitoring indicators have long-term monitoring data, they cannot be used, due to the lack of corresponding unit price. It is suggested that the methodology of biodiversity assessments should be developed in the future, to identify the corresponding indicators, and the scope of application and corresponding prices, so that the accuracy of the biodiversity value assessment can be increased.

4.4. Improve the Estimation of Soil Erosion

Regarding soil erosion prevention in the forest, this study, inhibited by time and funding, used RUSLE to calculate the difference in the amount of soil loss in the two scenarios of forest land and bare land, with each working circle of national forest as the scope. Then, we converted this into the amount of soil erosion in each working circle through SDR, assuming that there was forest land coverage in the working circle. However, a working circle can contain many watershed units, and the slope, surface coverage, and SDR of each watershed unit were bound to be different. Accordingly, to obtain a more accurate and detailed amount of soil, it is recommended to conduct a detailed investigation and calculation of the amount of soil erosion in individual cases for the main watershed units in different regions. In addition, the general formula calculated only the loss of surface soil, excluding the deep stratum and landslide. Normally, forest land has a small amount of runoff and soil erosion, but it is prone to collapse in extreme events, which can cause a large amount of soil erosion. The occurrence of landslides is atypical, and therefore this study did not include them in the estimation, but it can be further evaluated in the future.

4.5. Eosystem Disservices (EDS)

Ecosystem services are the benefits people obtain from ecosystems [2]. Research on the contribution of ecosystem services to human well-being has increased exponentially since MA. In contrast, some ecosystem disservices that undermine, or harm human well-being have been seriously overlooked [66]. The reason for this neglect of ecosystem disservices is that the frequency and intensity of their occurrence are variable (such as pest outbreaks) [67], and they lack clearly defined functions and types [68]. In Taiwan, substantial negative impacts, such as the landslide damage caused by the typhoon, the damage caused by the invasion of foreign species, and the protected Formosan macaques endangering crops, and other ecosystem disservices, are not included in the value estimation at present. To reflect the true ecosystem service value, EDS should also be reflected in the future, so that humans can better manage the ecosystem.

5. Conclusions

The forest is the largest ecosystem in the land area of Taiwan. In the past, most evaluation studies on the forest ecosystem services were regional, or focused on services such as conservation, soil conservation, and carbon sequestration, which lacked a representative evaluation of the value of forest ecosystem services of Taiwan. The evaluation of service values conducted by this study is on a nationwide forest ecosystem in Taiwan. The evaluation items included forest water conservation, air purification, biodiversity, soil erosion prevention, forest carbon storage, and forest recreation (including environmental education). It is preliminarily estimated that the total value of the forest ecosystem services in 2016 was approximately NT $749,278 million, accounting for roughly 4.28% of the GDP of the year.

The value of the forest ecosystem services in this study were derived by multiplying the size of each ecosystem service, defined in biophysical metrics, by its unit value. The evaluation results shed light on the link between ecosystem services and forestry management and conservation effectiveness. However, the gathered data on the amount of forest air pollution removal, the amount of soil erosion prevention and control, and on biodiversity indicators are not complete, and further research is still needed. Moreover, it is paramount to give proper scrutiny when drawing upon unit price data, for this has a huge impact on the final assessment results of forest ecosystem service values.

Author Contributions: Conceptualization, J.-C.L.; methodology, C.-R.C.; data curation, W.-H.C.; writing—original draft, M.-S.W. All authors have read and agreed to the published version of the manuscript.

Funding: This research was funded by Taiwan Forestry Research Institute, grant number 108AS-10.2.1-FI-G2", Forestry Bureau, grant number tfbc-1070206, and the APC was funded by Taiwan Forestry Research Institute.

Institutional Review Board Statement: Not applicable.

Informed Consent Statement: Not applicable.

Data Availability Statement: The data presented in this study are available on request from the corresponding author.

Conflicts of Interest: The authors declare no conflict of interest.

References

1. Gómez-Baggethun, E.; de Groot, R.; Lomas, P.L.; Montes, C. The History of Ecosystem Services in Economic Theory and Practice: From Early Notions to Markets and Payment Schemes. *Ecol. Econ.* **2010**, *69*, 1209–1218. [CrossRef]
2. Reid, W.; Mooney, H.; Cropper, A.; Capistrano, D.; Carpenter, S.; Chopra, K. *Millennium Ecosystem Assessment. Ecosystems and Human Well-Being: Synthesis*; Island Press: Washington, DC, USA, 2005.
3. Sukhdev, P.; Wittmer, H.; Schröter-Schlaack, C.; Neßhöver, C.; Bishop, J.; Ten Brink, P.; Gundimeda, H.; Kumar, P.; Simmons, B. *Mainstreaming the Economics of Nature: A Synthesis of the Approach, Conclusions and Recommendations of TEEB*; UNEP: Nairobi, Kenya, 2010.
4. Haines-Young, R.; Potschin, M. *CICES V4.3—Revised Report Prepared Following Consultation on CICES Version 4, August–December 2012*; EEA Framework Contract No EEA/IEA/09/003; University of Nottingham: Nottingham, UK, 2013.
5. United Nations. System of Environmental-Economic Accounting—Ecosystem Accounting (SEEA EA). 2021. Available online: https://seea.un.org/sites/seea.un.org/files/documents/EA/seea_ea_white_cover_final.pdf (accessed on 19 November 2021).
6. WAVES Partnership. *WAVES Annual Report 2015*; World Bank Group: Washington, DC, USA, 2015.
7. Tallis, H.; Kareiva, P.; Marvier, M.; Chang, A. An Ecosystem Services Framework to Support Both Practical Conservation and Economic Development. *Proc. Natl. Acad. Sci. USA* **2008**, *105*, 9457–9464. [CrossRef]
8. Martín-López, B.; Gómez-Baggethun, E.; García-Llorente, M.; Montes, C. Trade-offs Across Value-domains in Ecosystem Services Assessment. *Ecol. Indic.* **2014**, *37*, 220–228. [CrossRef]
9. Croci, E.; Lucchitta, B.; Penati, T. Valuing Ecosystem Services at the Urban Level: A Critical Review. *Sustainability* **2021**, *13*, 1129. [CrossRef]

10. Pascual, U.; Muradian, R.; Brander, L.; Gómez-Baggethun, E.; Martín-López, B.; Verma, M.; Armsworth, P.; Christie, M.; Cornelissen, H.; Eppink, F. *The Economics of Valuing Ecosystem Services and Biodiversity. TEEB-Ecological and Economic Foundation: Ecological and Economic Foundations*; Kumar, P., Ed.; Earthscan: London, UK, 2010.
11. De Groot, R.; Brander, L.; Van der Ploeg, S.; Costanza, R.; Bernard, F.; Braat, L.; Christie, M.; Crossman, N.; Ghermandi, A.; Hein, L.; et al. Global Estimates of the Value of Ecosystems and Their Services in Monetary Units. *Ecosyst. Serv.* **2012**, *1*, 50–61. [CrossRef]
12. Kubiszewski, I.; Costanza, R.; Dorji, P.; Thoennes, P.; Tshering, K. An Initial Estimate of the Value of Ecosystem Services in Bhutan. *Ecosyst. Serv.* **2013**, *3*, e11–e21. [CrossRef]
13. Costanza, R.; De Groot, R.; Braat, L.; Kubiszewski, I.; Fioramonti, L.; Sutton, P.; Farber, S.; Grasso, M. Twenty Years of Ecosystem Services: How Far Have We Come and How Far Do We Still Need To Go? *Ecosyst. Serv.* **2017**, *28*, 1–16. [CrossRef]
14. De Groot, R.S.; Wilson, M.A.; Boumans, R.M.J. A Typology for the Classification, Description and Valuation of Ecosystem Functions, Goods and Services. *Ecol. Econ.* **2002**, *41*, 393–408. [CrossRef]
15. Chan, K.M.A.; Satterfield, T.; Goldstein, J. Rethinking Ecosystem Services to Better Address and Navigate Cultural Values. *Ecol. Econ.* **2012**, *74*, 8–18. [CrossRef]
16. Bagstad, K.J.; Semmens, D.J.; Waage, S.; Winthrop, R. A Comparative Assessment of Decision-support Tools for Ecosystem Services Quantification and Valuation. *Ecosyst. Serv.* **2013**, *5*, 27–39. [CrossRef]
17. Costanza, R.; d'Arge, R.; Groot, R.; Farber, S.; Grasso, M.; Hannon, B.; van den Belt, M. The Value of the World's Ecosystem Services and Natural Capital. *Nature* **1997**, *387*, 253–260. [CrossRef]
18. Costanza, R.; de Groot, R.; Sutton, P.; van der Ploeg, S.; Anderson, S.J.; Kubiszewski, I.; Farber, S.; Turner, R.K. Changes in the Global Value of Ecosystem Services. *Glob. Environ. Chang.* **2014**, *26*, 152–158. [CrossRef]
19. Mengist, W.; Soromessa, T. Assessment of Forest Ecosystem Service Research Trends and Methodological Approaches at Global Level: A Meta-analysis. *Environ. Syst. Res.* **2019**, *8*, 22. [CrossRef]
20. Wang, B.; Ren, X.X.; Hu, W. China's Forest Ecosystem Service Function and its Value Evaluation. *For. Sci.* **2011**, *47*, 145–153.
21. Mitsubishi Research Institute, Inc. *Research Report on Evaluation of Multifaceted Functions of Agriculture and Forests Related to Global Environment and Human Life*; Mitsubishi Research Institute, Inc.: Tokyo, Japan, 2001.
22. Kim, J.H.; Kim, R.H.; Youn, H.J.; Lee, S.W.; Choi, H.T.; Kim, J.J.; Park, C.R.; Kim, K.D. Valuation of Nonmarket Forest Resources. *Korean J. For. Rec.* **2012**, *16*, 9–18. [CrossRef]
23. Lu, M.L.; Huang, J.Y.; Chung, Y.L. Assessing the Ecosystem Service Values in the Hengchun Peninsula, Taiwan. *Q. J. Chin. For.* **2012**, *45*, 491–501.
24. Wu, C.S.; Liu, C.P.; Chen, Y.H.; Chen, L.C.; Lin, J.C.; Jeng, M.R.; Hsu, C.Y. Evaluating the Economic Benefits of Forest Ecosystem Management in the LiuKuei Experimental Forest. *Taiwan J. For. Sci.* **2006**, *21*, 191–203.
25. Hsieh, C.H.; Liu, W.Y. Evaluating the Economic Value of Water Conservation by Taiwan National Forests. *Taiwan J. Appl. Econ.* **2018**, *104*, 185–228.
26. Taiwan Botanic Red List Editorial Committee. *The Red List of Vascular Plants of Taiwan, 2017*; Endemic Species Research Institute, Council of Agriculture, Executive Yuan, R.O.C (Taiwan): Nantou, Taiwan, 2017.
27. Accounting Office of Forestry Bureau, Council of Agriculture, Executive Yuan. *Forestry Statistics, 2017*; Forestry Bureau, Council of Agriculture, Executive Yuan, R.O.C (Taiwan): Taipei, Taiwan, 2017.
28. Martín-López, B.; Gómez-Baggethun, E.; González, J.A.; Lomas, P.L.; Montes, C. The Assessment of Ecosystem Services Provided By Biodiversity: Re-Thinking Concepts and Research Needs. In *Handbook of Nature Conservation: Global, Environmental and Economic Issues*; Aronoff, J., Ed.; Nova Science Publisher: New York, NY, USA, 2009; pp. 261–282. ISBN 978-1-60692-993-3.
29. Liu, S.; Costanza, R.; Troy, A.; D'Aagostino, J.D.; Mates, W. Valuing New Jersey's Ecosystem Services and Natural Capital: A Spatially Explicit Benefit Transfer Approach. *Environ. Manag.* **2010**, *45*, 1271–1285. [CrossRef]
30. Wilson, M.A.; Hoehn, J.P. Valuing Environmental Goods and Services Using Benefit Transfer: The State-of-the-art and Science. *Ecol. Econ.* **2006**, *60*, 335–342. [CrossRef]
31. Ninan, K.N.; Inoue, M. Valuing Forest Ecosystem Services- Case Study of a Forest Reserve in Japan. In *Valuing Ecosystem Services*; Edward Elgar Publishing: Cheltenham, UK, 2013; Volume 5, pp. 78–87.
32. National Forestry and Grass Administration. *Forestry Industry Standard of the People's Republic of China*; National Forestry and Grass Administration: Beijing, China, 2008; Volume 11.
33. Chen, H.H.; Li, J.Y. An Integrated Model of Soil and Water Conservation Functions in Forest Catchment Areas. *Q. J. Chin. For.* **1986**, *19*, 59–74.
34. Zheng, C.L. *Study on Water Use Functions and Water Use Pattern Prediction in Manufacturing Industry (1): Research Commissioned by Water Resources Agency*; Chung-Hua Institution for Economic Research: Taipei, Taiwan, 1994; Volume 13.
35. Wu, C.S.; Chen, Y.H.; Cheng, M.R.; Huang, J.L.; Lee, K.C. Evaluating the Economic Benefit of Water Resource Nourishment by Forests. *Taiwan J. For. Sci.* **2004**, *19*, 187–197.
36. LotSoar Consultants Co., Ltd. *The Study of Optimal Water Supply Portfolio for Central Region of Taiwan under the Impact of Climate Change*; Water Resources Agency, Ministry of Economic Affairs: Taipei, Taiwan, 2018; ISBN 9789860578973.
37. Nowak, D.J.; Crane, D.E.; Stevens, J.C. Air Pollution Removal by Urban Trees and Shrubs in the United States. *Urban For. Urban Green.* **2006**, *4*, 115–123. [CrossRef]

38. Xing, Y.; Brimblecombe, P. Role of Vegetation in Deposition and Dispersion of Air Pollution in Urban Parks. *Atmos. Environ.* **2019**, *201*, 73–83. [CrossRef]
39. Balestrini, R.; Tagliaferri, A. Atmospheric Deposition and Canopy Exchange Processes in Alpine Forest Ecosystems (Northern Italy). *Atmos. Environ.* **2001**, *35*, 6421–6433. [CrossRef]
40. Chen, Q.P.; Liu, C.P. The purification function of air pollutants by flat afforestation- with Pingtung Linhousilin Forest Park as an example. *For. Res. Newsl.* **2014**, *21*, 6–10. Available online: https://www.airitilibrary.com/Publication/alDetailedMesh?DocID=16056922-201412-201501120037-201501120037-6-10 (accessed on 17 March 2021).
41. Chen, Q.P. Air Pollutants Interception of Plains Afforestation in Central Taiwan. Ph.D. Thesis, Chung Hsing University, Taichung, Taiwan, 2015.
42. Su, Z.H.; Chen, Q.P.; Liu, C.P. Evaluation on the Purification Function of Air Pollutants by Tree Species on Flat Land. In Proceedings of the 2016 Flat Land Afforestation Test and Monitoring Seminar, Taipei, Taiwan, 22 December 2016; pp. 17–22.
43. Directorate General of Budget, Accounting and Statistics, Executive Yuan. *Green National Income Account of 2019*; Directorate General of Budget, Accounting and Statistics, Executive Yuan: Taipei, Taiwan, 2019.
44. Ni, P.C.; Lee, C.Y.; Yeh, F.L. Study of Air Quality for Taiwan Area. Available online: http://tasder.org.tw/meeting/2004/part01/1-01.pdf (accessed on 26 September 2017).
45. National Statistics. Domestic Production and Deflator Index of Various Industries over the Years. Taipei, Taiwan. Available online: https://www.stat.gov.tw/ct.asp?xItem=37407&CtNode=3564&mp=4 (accessed on 13 September 2017).
46. Butchart, S.H.; Walpole, M.; Collen, B.; Van Strien, A.; Scharlemann, J.P.; Almond, R.E.; Baillie, J.E.; Bomhard, B.; Brown, C.; Bruno, J.; et al. Global Biodiversity: Indicators of Recent Declines. *Science* **2010**, *328*, 1164–1168. [CrossRef]
47. Taiwan Biodiversity Observation Network (TaiBON). Available online: https://taibon.tw/en/node/20 (accessed on 25 September 2017).
48. Chen, Z.Y.; Zhang, H.Y.; Guan, L.H.; Zheng, H.Y. *Analysis of Economic Benefits of Nature Reserves. Agricultural Policy and Agricultural Conditions*; Council of Agriculture, Executive Yuan: Taipei, Taiwan, 2012; Volume 240, pp. 77–80. Available online: https://www.coa.gov.tw/ws.php?id=2445783 (accessed on 12 December 2020).
49. Renard, K.G.; Foster, G.R.; Weesies, G.A.; Porter, J.P. RUSLE: Revised Universal Soil Loss Equation. *J. Soil Water Conserv.* **1991**, *46*, 30–33.
50. Environmental Protection Administration, Executive Yuan. *The Republic of China (Taiwan) National Greenhouse Gas Inventory Report of 2018*; Environmental Protection Administration, Executive Yuan: Taipei, Taiwan, 2018.
51. European Bank for Reconstruction and Development. Methodology for the Assessment of Coal Fired Generation Projects. 2014. Available online: https://www.ebrd.com/documents/climate-finance/methodology-for-the-assessment-of-coal-fired-generation-projects.pdf (accessed on 2 February 2021).
52. Tourism Bureau, Ministry of Transportation and Communication. Survey of Travel by R.O.C. Citizens. 2016. Available online: https://admin.taiwan.net.tw/FileUploadCategoryListC003340.aspx?CategoryID=7b8dffa9-3b9c-4b18-bf05-0ab402789d59 (accessed on 10 May 2017).
53. Chang, H.H.; Chen, C.H. The Study of Visitor's Willingness to Pay for Interpretative Service and Environmental Conservation: A Case of FuShan Nature Preserve Park. *J. Tour. Travel Res.* **2010**, *5*, 57–76.
54. Tseng, C.H.; Lee, M.T. A Case Study of Tourist Willingness to Pay for Guidance Services: The Fushan Botanical Garden. *J. Commer. Mod.* **2008**, *4*, 29–49.
55. Soil and Water Conservation Bureau of the Council of Agriculture, Executive Yuan. *Water and Soil Conservation Handbook*; Soil and Water Conservation Bureau of the Council of Agriculture, Executive Yuan: Nantou, Taiwan, 2005.
56. Soil and Water Conservation Bureau of the Council of Agriculture, Executive Yuan. *Reference Manual for Overall Investigation and Planning of Watershed Area*; Soil and Water Conservation Bureau of the Council of Agriculture, Executive Yuan: Nantou, Taiwan, 2008.
57. King, H.B.; Liu, C.P.; Hsia, Y.J.; Huang, J.L. Interactions of the Fushan Hardwood Forest Ecosystem and the Water Chemistry of Precipitation. *Taiwan J. For. Sci.* **2003**, *18*, 363–373.
58. Zhang, B.; Li, W.; Xie, G.; Xiao, Y. Water Conservation of Forest Ecosystem in Beijing and Its Value. *Ecol. Econ.* **2010**, *69*, 1416–1426.
59. State of the Voluntary Carbon Markets. 2017. Available online: https://www.cbd.int/financial/2017docs/carbonmarket2017.pdf (accessed on 10 May 2017).
60. Dasgupta, P.S. The Welfare Economic Theory of Green National Accounts. *Environ. Resour. Econ.* **2009**, *42*, 3–38. [CrossRef]
61. Obst, C.; Hein, L.; Edens, B. National Accounting and the Valuation of Ecosystem Assets and Their Services. *Environ. Resour. Econ.* **2016**, *64*, 1–23. [CrossRef]
62. Mace, G.M.; Norris, K.; Fitter, A.H. Biodiversity and Ecosystem Services: A Multilayered Relationship. *Trends Ecol. Evol.* **2012**, *27*, 19–26. [CrossRef]
63. Wang, B.; Zheng, Q.H.; Guo, H. Economic Value Assessment of Forest Species Diversity Conservation in China Based on the Shannon-Wiener index. *For. Sci. Res.* **2008**, *21*, 268–274.
64. Zhang, Y. Research on the Value Assessment of Forest Biodiversity in China. *For. Econ.* **2001**, 3, 37–42.
65. Choi, H.A.; Lee, W.K.; Song, C.; Forsell, N.; Jeon, S.; Kim, J.S.; Kim, S.R. Selecting and Applying Quantification Models for Ecosystem Services to Forest Ecosystems in South Korea. *J. For. Res.* **2016**, *27*, 373–1384. [CrossRef]

66. Ninan, K.N.; Inoue, M. Valuing Forest Ecosystem Services: What We Know and What We Don't. In *Valuing Ecosystem Services*; Edward Elgar Publishing: Cheltenham, UK, 2013; Volume 93, pp. 137–149.
67. Lyytimäki, J.; Sipilä, M. Hopping on One Leg: The Challenge of Ecosystem Disservices for Urban Green Management. *Urban. For. Urban Green.* **2009**, *8*, 309–315. [CrossRef]
68. Shackleton, C.M.; Ruwanza, S.; Sanni, G.S.; Bennett, S.; De Lacy, P.; Modipa, R.; Mtati, N.; Sachikonye, M.; Thondhlana, G. Unpacking Pandora's Box: Understanding and Categorising Ecosystem Disservices for Environmental Management and Human Well-being. *Ecosystems* **2016**, *19*, 587–600. [CrossRef]

MDPI

Article

On the Mismatches between the Monetary and Social Values of Air Purification in the Colombian Andean Region: A Case Study

Andres Suarez [1,*], Cesar Ruiz-Agudelo [2], Edisson Castro-Escobar [3], Gloria Y. Flórez-Yepes [4] and Luis A. Vargas-Marín [3]

1 Department of Civil and Environmental, Universidad de la Costa, Calle 58#55-66, Barranquilla 080001, Colombia
2 Doctoral Program in Environmental Sciences and Sustainability, Universidad Jorge Tadeo Lozano, Bogota 111311, Colombia; cesara.ruiza@utadeo.edu.co
3 Environment and Development Research Center (CIMAD), Universidad de Manizales, Carrera 9a #19-03 B/Campo Hermoso, Manizales 170001, Colombia; ecastro@umanizales.edu.co (E.C.-E.); lvargas@umanizales.edu.co (L.A.V.-M.)
4 GIDTA Research Group, Universidad Católica de Manizales, Manizales 170001, Colombia; gyflorez@ucm.edu.co
* Correspondence: asuarez24@cuc.edu.co

Citation: Suarez, A.; Ruiz-Agudelo, C.; Castro-Escobar, E.; Flórez-Yepes, G.Y.; Vargas-Marín, L.A. On the Mismatches between the Monetary and Social Values of Air Purification in the Colombian Andean Region: A Case Study. *Forests* **2021**, *12*, 1274. https://doi.org/10.3390/f12091274

Academic Editors: Elisabetta Salvatori and Giacomo Pallante

Received: 9 August 2021
Accepted: 15 September 2021
Published: 17 September 2021

Abstract: There is growing interest in air quality and air purification, due to current high pollution levels, their effects on human health, and implications for urban economies. Since the improvement of air quality carries important economic value, air-related benefits have been evaluated monetarily from two perspectives: the first relates to air quality improvements, while the second values air purification as an ecosystem function. This research opted for the second perspective, given that the study area (two Colombian municipalities) does not suffer from poor air quality conditions, but stakeholders prioritized this function as highly important to them. Contingent valuation methods were applied in order to identify the population's probability of willingness to maintain the air purification function. Although individuals (n = 245) attribute a yearly monetary value of USD 1.5 million to air purification, it was found that, despite the high level of social importance that respondents assigned to air purification (mean = 4.7/5), this had no correlation with payment values (rho = 0.0134, p = 0.8350); that is, households do not really recognize the monetary value of all the benefits they receive or the benefits they would lose if the service suffers changes. Hence, it is posed that monetary values do not necessarily reflect the social importance that individuals assign to ecosystem services, and attention is called to the need to integrate social and monetary values into decision-making processes, so as to encompass the complexity of ecosystem services and conciliate conflicting valuation language.

Keywords: air purification; deliberation; forest ecosystems; economic valuation; social valuation

1. Introduction

It is well known that humankind's well-being relies on different ecosystems and their benefits, on multiple scales, particularly in urban areas. In this regard, ecosystem services (ESs) have gained momentum, as this approach is insightful for decision-making purposes. According to [1], ESs are those ecosystem aspects that lead to human wellbeing. Thus, the generation of ESs begins with the presence of one or more ecosystem structures, which have a variety of ecological functions, and ultimately provide societal benefits [2,3]. As stated by [4], the ecosystem services approach enabled a better understanding of the complexity of human-environmental systems. As stressed by [5], it is necessary to manage ecosystem functions in order to ensure that the composition and structure of ecosystem elements continuously provide well-being for humanity. There are certain ecosystem functions that are essential to the maintenance of said well-being, which makes the case for the maintenance of good environmental conditions for these reasons, including air quality, which has received special attention in the past few years [6,7].

Given their role in carbon sequestration, it may be established that forest ecosystems contribute to air purification [8,9]. They are also very important, given their capacity for the absorption of atmospheric contamination [10–13]. The air purification (AP) function provided by vegetation is often quantified as an annual removal of pollutants, mainly on the urban scale [14]. In general, AP has been understood as a human welfare indicator, in terms of air quality or clean air [15,16]. AP mainly embodies the production of negative tropospheric ozone, absorption of sulfur dioxide, nitrogen oxides, fluoride, particulate matter (PM10), and noise reduction [17,18]. Further, they contribute to a final benefit related to clean-air environments, by removing pollutants from the atmosphere, reducing the cost of human-led air purification, and preventing large expenditures on public health and safety [19].

As the purification of air has important economic values, a common valuation framework has historically been the contingent valuation method [20–23]. Air-related benefits have been evaluated from two perspectives: that related to air quality improvements, and the air purification function, per se, which helps to identify people's willingness to pay (WTP) for air-related processes [24–29]. However, the analysis of the social and monetary values of AP, as a process previous to air quality improvements, has received little attention.

Human decisions and behaviors regarding and toward ecosystems are also determined by the multiple ways in which nature, ecosystems, or ecosystem services are important for individuals or social groups [30–33]. Especially, in the scientific field of ecosystem services, monetary valuation methods have received more attention than other valuation methods [34]. Concentrating only on monetary valuation underscores instrumental values while ignoring both intrinsic and relational values [33]. In recent years, academics that specialize in ecosystem service valuation have adhered to value pluralism: the recognition of different, and often conflicting, value domains that are neither reducible to each other nor to some ultimate value [34–36].

Therefore, this study did not analyze WTP for air quality improvements or strategies aimed at the removal of air pollution from urban areas but instead obtained an ES valuation linked to the surrounding forest ecosystems of two urban areas. These ecosystems promote air quality through their air purification function. The social prioritization of AP supports this valuation, as individuals regard the air purification process as an essential forest function for potential AP, even if there are favorable atmospheric conditions in their areas. In this sense, it is essential to preserve these conditions for the future.

Therefore, the present study aims to contrast the ways in which AP is valued from different perspectives, in order to understand local valuation approach mismatches. To that end, a social and monetary valuation regarding AP, provided by the main ecological structure in two Colombian municipalities, was performed. This main ecological structure consists of those natural and semi-natural areas that promote ecosystem services and are involved in legal Colombian land use planning.

Methods for Monetary Valuation

Air valuation is a necessary resource for the sustainability of the human species. It does not present a defined market, much less does it present costs for its generation, because this does not depend on human effort. However, the alteration of its structure produces negative effects on humans. Certain valuation methodologies consider value from a cost perspective; thus, "This implies that air pollution (alteration of the minimum levels of acceptance by the human being) must be valued at the cost of correcting the emerging damage" [37], or from other approaches, such as replacement or corrective cost.

Although the choice experiments method has been highlighted as a powerful technique for the prevention of biases and improvement of valuation scenarios, through marginal effects identification, given the complexity of AP and the cognitive burden linked to these experiments, this investigation opted to use the contingent valuation (CV) method. Contingent valuation could be used in scenarios in which the generation of ESs could be explained from a broader perspective, rather than the provision of confusing attributes specific to local communities' valuation, particularly for AP. For instance, authors such as [12] call attention to the need to analyze AP through the assessment of pollutant sequestration, specifically SO_2 and NO_2. Moreover, the authors of [38] suggest approaches such as functional diversity, for pollution absorption. With this in mind, the application of choice experiments could configure incomprehensible attributes for the inhabitants of the municipalities studied herein. In the scientific literature, the combination of match air purification AND choice experiments is scarce. There is a single study, performed in [16], which provides a framework for the application of a choice experiment in AP. However, the investigation was limited to the provision of only those attributes related to increasing area (versus a status quo area). In the context of the present study, this approach was unreliable because the current context of the study area seeks the promotion and conservation of current areas, in order to declare them the main ecological structure.

Therefore, the ES monetary valuation was performed through a contingent valuation method. This method is useful, since it is flexible and could be applicable to many non-marketed and marketed ESs and allows for the capture of all types of benefits from an ES, including instrumental and non-instrumental values [39]. However, it is known that this method has several weaknesses, such as its high level of sensitivity to survey application, and the potentiality for study biases, which could limit result generalizability [40]. Despite these limitations, contingent valuation has allowed, in different contexts, for the recognition of relationships between the willingness to pay given a population's characteristic, and the opportunity cost of protecting ecologically fragile areas vulnerable to degradation [41].

As defined by [42], the possibility of direct interviews, with contextual explanations of these opportunity costs, permits the limitation of regressive biases and the adverse selection of contingent valuation. The key lies in the contextualization made by researchers about the valuation scenario, such that interviews conducted by indirect means are usually more questioned, as they tend to induce biases and greater response subjectivity [43].

2. Materials and Methods

2.1. Research Area

This study was carried out in two Colombian municipalities, located in the northern region of the Department of Caldas (Figure 1). These are mainly rural areas and possess less than 31 thousand inhabitants between them (Aguadas = 21,043, Pácora = 10,608). The current regional landscape has been largely transformed (approximately 20% of the original ecosystems remain) by human interventions in the last 350 years [44]. Nowadays, a huge portion of the land's surface is used for agriculture, mainly to grow coffee and avocados, raise cattle and grasslands. Said activities have led to the loss of forests. It is noteworthy that only one protected area exists in the above-mentioned zone: the protected "Tarcará", a forest reserve located between the two municipalities under study. In order to manage area ecosystems, the local environmental authority, Corpocaldas, supported the identification, valuation, and projection of the main ecosystems in the zone (main ecological structure), aiming to sustain the generation of ESs for land use planning. In this context, this study was performed to provide information about ES valuation mismatches for local managers, and to subsequently shed light on the need to approach ecosystem management from the perspective of both simplistic (exclusively monetary valuation) and pluralistic valuations.

2.2. Ecosystem Service Prioritization and Valuation

A comprehensive approach was proposed to elucidate the values of ESs in the study area. In the present research, this was performed via a two-level analysis—one related to ES deliberative prioritization by stakeholders and an individual-based valuation by way of a survey with a local population sample (valuation of the ESs prioritized).

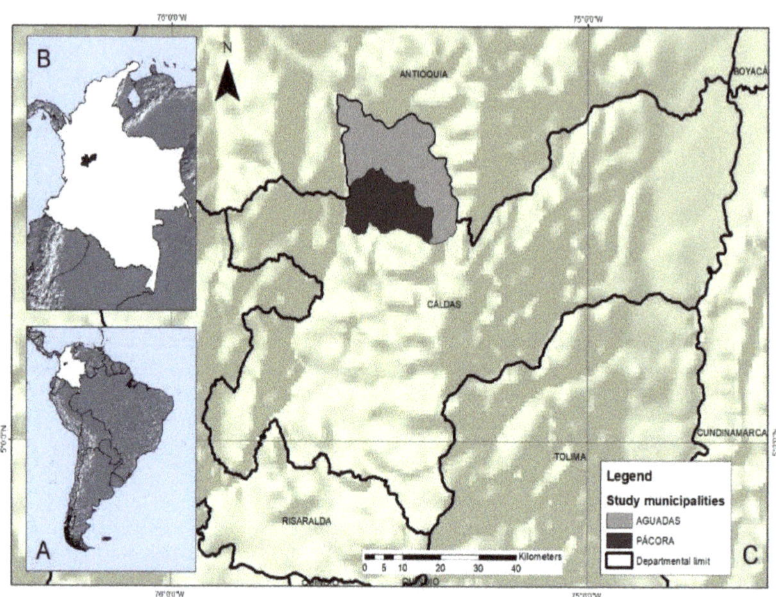

Figure 1. Location of the study area: (**A**) Latin America; (**B**) Colombia and Caldas; (**C**) Aguadas and Pácora.

2.2.1. Deliberative Prioritization Process

Two focus groups were convened (one per municipality) to discuss the most important ecosystem services (ESs) for different stakeholders. The recruitment process for the focus group's participants was as follows: first, the local environmental authority provided a list of municipal stakeholders (laypeople, academics, civil servants). Later, calls were made to recruit as many local players as possible to represent different area views. From the municipalities of Aguadas and Pácora, 11 and 26 people participated in the focus groups, respectively, where a deliberative approach to identify ESs (3–4 h discussions, reflections, and deliberation to prioritize ESs) was employed. Researchers provided refreshments, and when needed, attendee transportation expenditures were assumed for some of Pacora's participants.

The sessions occurred in accordance with the steps below:

i. The participants were divided into subgroups to discuss and debate, more extensively, the ESs used and their perceptions thereof.

ii. Researchers provided formats to guide internal discussions. The guide form specified that participants should rank, between 1–5 on the importance scale, those ecosystem benefits that were most important to them, first as individuals, and then as a subgroup (Table S1: Main ecological structure in the two municipalities). In this step of prioritization, participants were encouraged to discuss the most important benefit for them. Additionally, a participatory mapping approach was employed, in which stakeholders were invited to spatialize the priority areas for ES supply. A map of the municipality was provided, with different forest covers. A final guide question was formulated: what kinds of benefits do you receive from the identified municipality´s main ecological structure? (Table S2. Ecosystem Services identification and prioritization).

iii. In the plenary session, participants consolidated a list of the most important ESs for their livelihoods (daily life, economic and social development, and ecosystem resilience). The identifications and prioritizations were based on participants' experience and preexisting knowledge of ecosystems in said municipalities and their benefits. In both municipalities, air purification (AP) was prioritized as one of the most important benefits provided by forests.

Although each municipality prioritized five ESs, the decision for the selection of air purification for this document was made because it was a common prioritized benefit between the municipalities. Although AP could be considered a function rather than a final ES, stakeholders prioritized AP function as highly important to them, and as a precursor for posterior benefits, including future air quality.

2.2.2. Contingent Valuation Methodological Process

A questionnaire was designed to elicit information about socioeconomic characteristics, perceptions about ESs (first prioritized in workshops by stakeholders), and a section relating to their willingness to pay (WTP) for AP was completed (Table 1). The questionnaire underwent an initial adjustment process, through group discussions, with professionals in environmental engineering, biology, and environmental management, as well as with laypeople. The final version of the instrument was built, and in connection with the WTP scenario, an open-ended format was proposed for two reasons: the municipalities' lack of initial reference information with which to generate values and payment ranges, and the current literature that justifies the ratio of open-ended questions in this type of scenario [8,45]. As the elicitation mechanism did not provide a predetermined amount to be paid, all other aspects of the payment were described, such as who pays, the mechanism for said payment, and temporality [46]. A purposive sampling to select sampling points that covered different urban zones was proposed. Urban sampling points encompassed neighborhoods from both Aguadas and Pácora—95% confidence and 9% error values were applied.

Table 1. Summary of the contingent valuation methodological process.

Step	Definition
Definition of ES change	Main ecological structure maintenance to sustain air purification function.
Specification of the contingent valuation model, definition of variables, indicators, and required data	Definition of binary model WTP = Yes (1), WTP = No (0). Application of the Logit model to identify the probability of responding Yes, and Tobit model to obtain the probability of the maximum value to be paid (censored model).
Specification of the measure for the change in welfare to be estimated	$E(WTP \mid X's) = \beta 0 + \beta VS$ (1) Where E is the expected value of the estimated willingness to pay, with data from the open-ended format, beta parameters are the estimated regression coefficients, vs. is a vector of socioeconomic variables and interviewee perception.
Design of the data collection instrument: survey	Designing questions about socioeconomic characteristics, perceptions about the ESs prioritized in the stage of work with stakeholders, and the section related to the WTP.
Sample size calculation and choice of sampling approach	Simple random sampling with 95% confidence and 9% error. Aguadas (*N* = 21,043), Pácora (*N* = 10,608).
Final survey application	Face-to-face meeting in each municipality.
Estimation of the econometric model	Logit and Tobit application in Stata 14®

Given the CV consideration mentioned, in this study, the threats and importance of ecological structure for inhabitants, with assisted questions in an interactive dialogue, were explained to respondents. Additionally, to prevent biases, the ex-post calibration techniques of [47] were followed. For example, researchers (i) emphasized the budgetary consequences of the WTP scenario and respondent choices, (ii) researchers urged respondents to be honest and act as if they had to act at the moment, and (iii) the study sought to reduce the bias of social desirability (the tendency to answer that which the interviewee considers socially acceptable) by mentioning some unrealistic examples from previous surveys.

The WTP scenario was as follows: First, trends in the main ecological structure, in both municipalities were explained. These included increases in crop production and growing homogenous land cover (mainly avocado crops) which have led to ecosystem degradation. Later, the study proposed a scenario in which the current main ecological structure was managed, and the local authorities promoted conservation strategies to avoid ecosystem loss and degradation. Finally, the WTP proposed question was "How much would you be willing to voluntarily pay to maintain the current main ecological structure in your municipality and to prevent future degradation"? The question was applied both in Aguadas and Pácora, in accordance with their particularities. Respondents were advised that, in this scenario, by carrying out the necessary conservation actions, they would need to make a monthly payment through utility bills (payments by household). This strategy allowed them to limit payments to within the constraints of household spending.

Regarding the empirical model involved in this exercise, two levels of analysis were used: the application of a Logit model, with marginal effects, by which to identify the determinants in respondents' willingness to pay, whereas the second one was a censored Tobit model (given the zero results for open WTP questions) with marginal effects for the identification of those factors that influenced maximum monetary values that respondents would pay.

Additionally, through the Mann-Whitney U-test, the values stated by respondents in the two municipalities ($p < 0.05$) were tested for differences among these. Models were calculated using the Stata 14® software (StataCorp, 4905 Lakeway Drive, College Station, TX, USA).

Finally, the questionnaire inquired about the importance (1–5 scale) that respondents attributed to the AP for monetary valuation purposes. Although WTP usually aimed to translate the random utility of an individual in monetary terms, we intentionally pointed out that although WTP demonstrates social utility, when asking for social perspectives apart from WTP, people assign even greater social values beyond what the payment represents.

Therefore, an open-ended question on the WTP for this ES was applied, and the following hypothesis was proposed to test instrumental and non-instrumental values: *there is a significant relationship between social importance and monetary values assigned to the prioritized ES*. This hypothesis ought to discover whether the higher the social importance of the ES, would indicate higher values that individuals would be willing to pay. To this end, a Spearman's correlation test ($p < 0.05$) was applied.

3. Results

3.1. Socioeconomic Characteristic of the Respondents

A total of 245 people, from both municipalities, were sampled. Most respondents were female (63%), and fell within the over-40 years-of-age range (51%), had middle or low education levels, and were mainly urban residents (93%). The respondents stated that they had low-income levels (53% reported earning less than one minimum Colombian legal wage), and most also stated that they were familiar with the natural ecosystems in their municipalities (84%). Table 2 shows socioeconomic information by municipality, detailing the frequencies and percentages of the items considered by this study.

Table 2. Respondent socioeconomic information (n = 245).

Municipality	Aguadas (n = 124)		Pácora (n = 121)		Consolidated (n = 245)	
Variable	Frequency	(%)	Frequency	(%)	Frequency	%
Gender						
Male	52	42	51	42	103	46
Female	72	58	70	58	142	63

Table 2. *Cont.*

Municipality	Aguadas (n = 124)		Pácora (n = 121)		Consolidated (n = 245)	
Variable	**Frequency**	**(%)**	**Frequency**	**(%)**	**Frequency**	**%**
Age						
15–17	3	2	12	10	15	7
18–25	25	20	23	19	48	21
26–39	35	28	33	27	68	30
>40	61	50	53	44	114	51
Education level						
None	2	2	5	4	7	3
Elementary	26	21	23	19	49	22
High school	49	39	73	60	122	54
Diploma Course	25	20	11	9	36	16
Undergraduate	18	14	8	7	26	12
Postgraduate	4	3	1	0.8	5	2
Residence						
Urban	110	89	107	88	217	97
Rural	14	11	14	12	28	3
Employed						
Yes	97	78	85	70	182	81
No	27	22	36	30	63	28
Incomes						
<1MCLW *	66	53	63	52	129	58
1–2 MCLW	42	34	45	37	87	39
>2 MCLW	15	12	13	11	28	13
Familiar with the ecosystems in their municipalities						
Yes	106	86	82	68	188	84
No	17	14	38	32	55	25

* 1 Minimum Colombian legal wage = COP 828,116; USD 1 = COP 3356 (01/24/20).

3.2. Monetary Valuation of EcosystemEcosystem Services

Air purification was an important ES for respondents (mean = 4.7, SD = 0.04). A total of 58% provided positive responses, in accordance with the WTP, while 42% stated a negative answer regarding WTP. This indicates that, while most individuals are willing to pay, a high percentage do not agree with the valuation scenario. Regarding the monetary value assigned to the ES, no significant difference was found among those amounts stated in the municipalities ($p = 0.8257$). As for the probability to express a positive WTP, the Logit model, with marginal effects, depicts those variables which influenced responses (Table 3).

Table 3 shows the results of the Logit model, with marginal effects. According to the results of the present investigation, the model was properly specified, in accordance with the errors ($p > 0.05$). It classified 70.83% of the answers, and in accordance with the Hosmer-Lemeshow test, the model's fit is appropriate ($p > 0.05$). It was found that respondents with higher levels of education and those who formed part of social organizations were more likely to present positive WTPs. Regarding the marginal effects of the Logit model, it was identified that higher schooling levels indicated 33% more likely to state positive WTPs than those with lower education levels. Additionally, forming part of a social organization contributed to a 25% higher likelihood to state positive WTPs (Table 3).

On the other hand, the maximum WTP was tested by the Tobit model (Table 4). In this case, the Tobit model produced estimators in accordance with WTP distribution, that is, to measure the social value for a change in the supply of AP. When an explained variable has a natural bias, truncated models balance estimators. The Tobit model showed that those under 40 were willing to pay COP 029 less than >40-year-old respondents. Additionally, in connection to the non-educational level, those who had been educated through elementary would pay COP 6062, and others would pay COP 12,329 if they had been educated through high school. Finally, forming part of a social organization increased the said contribution by COP 1353 (Table 4).

Table 3. Logit model with marginal effects.

WTP-Yes	Coefficient	S.E.	z	$p >$ |z|	dy/dx
Gender	0.367	0.346	1.060	0.289	0.3303
Age	−0.068	0.338	−0.200	0.841	−0.0138
Education level	1.622	0.316	5.130	0.000 *	0.3303
Residence	0.054	0.523	0.100	0.918	0.0110
Organización	1.216	0.686	1.770	0.076 **	0.2476
Knows	0.418	0.386	1.080	0.280	0.0851
Air importance	0.037	0.234	0.160	0.873	0.0076
_cons	−3.571	1.365	−2.620	0.009	
% Classification		70.8%			
Log likelihood =		−113.76162			
Prob > chi^2 =		0.0000			
Pseudo R^2 =		0.1435			
R2 McFadden		0.144			
Roc curve		0.7536			
Hosmer-Lemeshow chi2 =		0.3297			

Sig. * $p < 0.01$; ** $p < 0.10$.

Table 4. Tobit (censored).

WTP	Tobit
Gender	−6029.2
Age	−10,115.2 *
Education level	6062.6
Residence	2551.6
Organization	−1353.0
Knows	−1449.8
Air importance	4934.2
_cons	−14,967.6
/sigma	20,649.9
Number of obs =	91
left-censored WTP ≤ 0	1
Log likelihood =	−1022.7472
Prob > chi^2 =	0.1302
Pseudo R^2 =	0.0054
Prob > F =	0.0037
R^2 =	0.1066
Breusch-Pagan Prob > chi^2 =	0.0000
Test White Heteroskedasticity (p)	0.8064
Ramsey test Prob > F =	0.0012

Sig. * $p < 0.05$.

Finally, in terms of the monthly amount, it was more likely that those from the municipalities were willing to pay a total of 13,784 COP/month. The aggregated value for the entire population, in both municipalities, was 436,277,384 COP/month, or 5,235,328,608 COP/year (approximately 1.5 million USD/year).

3.3. *Social Valuation*

Regarding the stated hypothesis, no correlation was found between the importance level and the amount they were willing to pay (rho = 0.0134, p = 0.8350). In this sense, regardless of the high importance assigned to AP, there was no correlation with WTP values. In this case, higher levels of importance in the ES did not ensure higher payment amounts. It is important to highlight that, correlation does not necessarily mean causation, and thus, the fact that the importance of air as a variable is not causing the WTP is already represented by the insignificant parameter of the Logit model (Table 3). Although this study found conflicted valuation languages, it is important to stress that several value domains can coexist in the same area. This is so because of the instrumental values captured through a WTP, which only expressed utility measurements for respondents, while other values of AP, such as fundamental values (importance for sustaining life) and intrinsic values, were highlighted as non-instrumental relations with nature, as these value domains often express a sense of collective meaning [33].

4. Discussion

4.1. Importance and Monetary Valuation of Air Purification

Regarding the identification of AP as a priority, the literature emphasizes that AP is a highly valued ES in urban contexts (in terms of social importance level), given its contribution to urban quality of life [8,17,48]. Although the research on air quality recognizes the great importance of the value to be paid, few studies assign a specific amount to AP because these values are involved in multiple ES assessment contexts [8,22]. However, regarding the monetary value of AP, per se, a study carried out by [49] identified an amount of 12.27 USD/month, while [50] stressed that air purification was the main reason to pay a total value of 9.90 USD/month for green urban areas. In contrast, the present study found a total of 13,784 COP/month (USD 4.1), which is a significantly lower amount. One reason that explains the result is related to the high levels of negative WTP (42%). Such a high negative WTP means that, even if stakeholders prioritized air purification as a fundamental ES, and respondents rated AP with high values of importance, in general, they did not find "air quality issues" in their municipalities. Additionally, it means that some respondents are "adverse" to the proposed change and should be compensated. Conversely, in highly polluted urban areas, positive WTP reached 78% [26] even 84% [25], which demonstrates that AP valuation is context-specific.

Regarding WTP application via the contingent valuation method, it is important to note certain limitations, as highlighted by several authors [51]. Within the context of this investigation, results only showed pathways to the understanding of one scenario, rather than several potential scenarios, regarding environmental trade-offs and WTP particularly, within the context of environmental changes [52]. Hence, it is important to perform further research to deepen the understanding of several WTP aspects, in addition to social valuation implications.

4.2. Mismatches between Monetary Values and Social Values of Air Purification in the Present Case

This study shows that, regardless of the high importance assigned to AP, there was no correlation with WTP values. In this case, higher levels of importance in the ES did not ensure higher payment amounts. These results invite a deepening of the plural valuation, defined as a science-policy process that assesses the multiple values attributed to nature by social actors, and the ways in which this knowledge can guide decision making [36].

In this case, study, monetary (instrumental) values did not necessarily reflect the social importance that individuals assigned to ES, for two reasons: first, when stressing the importance of a prioritized ES "in abstraction" (i.e., theoretical importance given to the ES), the societal validation process could fit the importance of ES into a context-specific scenario (e.g., low air quality issues), which confronts social values with monetary values (i.e., not paying a lot for something that is already "good"). Secondly, as highlighted by many researchers, limiting ES valuation to monetary units generates insufficient approaches to ES complexity [48,52–55].

The values that individuals place on ecosystem services have been identified as a critical dimension of the sustainable management of social-ecological systems. In recent times, the call for the integration of plural ecosystem values, beyond just intrinsic and instrumental values, has inspired the notion of "relational values" [56]. Plural valuation approaches could be improved by distinguishing relational from instrumental values and by expressing these in nonmonetary terms. The results of the present study argue that multiple ecosystem values expressed by societies should be included in environmental management, so as to tackle social conflicts and consider the diverse needs and interests of different stakeholders.

In agreement with [57], in environmental economics, it is key to seek an adequate balance between disciplinary excellence, interdisciplinary collaboration, and political impact. In [58], the insufficiency of collaboration between economists and other disciplines is clearly demonstrated. With this value pluralism approach, scholars can integrate multiple disciplines, as well as qualitative and quantitative methods in ecosystem service valuation [37,59,60].

In this sense, although it is not always necessary to express the value of ESs in monetary terms, such values may be useful if the outcome delivers information to support decision making, and provide technical support, as well as information [61]. Recognizing the plural values that individuals attribute to ecosystems is a critical research priority toward the sustainable management of ecosystems. Pluralistic valuations may aid in aligning management interventions with individuals' values [30,31], as well as the identification of consensual and conflicting values connected with management approaches [30,31,62]. Therefore, it is strongly recommended to not consider only the 1.5 million USD/year value in the study area but the need to integrate social and monetary values into the perspective of concrete actions aimed to sustain forest areas for AP as well.

Therefore, any conventional method of microeconomic air valuation would lead to imprecise monetary quantifications, due to its classification as a public good without rivalry and excludability. The above implies that perception surveys, despite their basis on assumptions, provide a starting point for indirect assessment, in accordance with the sensitivity of the population and their degree of awareness of the environmental deterioration of air, to implement anthropic factors that communities have identified previously.

Numerous value articulating methods have been applied for ES valuation, from the viewpoint of the beneficiaries' subjective appreciation [63–66]. New developments include the integration of indigenous and local knowledge systems and practices [67], the development of integrative frameworks [68], and the comparative study of methods' capacities to capture plural values [33,60]. In order to address these issues, not only is there a need to develop proper assessment methods, but also to disclose the theoretical basics of this assessment, and which trade-offs go hand-in-hand with different assessment types [63–65]. Completely new issues for the valuation of ecosystem services and natural capital arise with the development of new tools.

5. Conclusions

Helping ecosystem services' monetary and social values to match is of great interest for both the academic and policy-making communities since this joint analysis would allow for the obtention of a broader understanding of the values of ESs in specific contexts. The present study thus aimed to perform a valuation of the air purification ES from a monetary and social perspective, in order to highlight ES valuation mismatches. Therefore, one conclusion from this investigation is that the economic valuation, expressed in monetary units, does not necessarily reflect the social importance attributed to the AP ecosystem service. It does indicate that, although the monetary value of AP was about 1.5 million USD/year in the study area, the results reflect that the statement "the higher the importance given to AP is, the higher the monetary value will be" was not suitable for the study area. Thus, it is of interest to integrate social and monetary values into the decision-making process, in order to encompass the complexity of ES and conciliate mismatches to generate more accurate socio-ecosystem management strategies, or to make decisions considering social perspectives (ES importance) and monetary values, so as to address valuation mismatches.

The mismatches between economic and social valuations reflected in this case study show that, in order to understand the "integral value" of ecosystem services, it is essential to move toward more pluralistic valuations, which contemplate instrumental values (such as economic valuation), but also sociocultural and intrinsic values. Understanding the ES value in real-life field exercises, such as that presented herein, highlights the immense sociocultural complexity and multiple views of value that stakeholders may have. A final recommendation for future research is undoubtedly to advance to more pluralistic evaluations.

Supplementary Materials: The following are available online at https://www.mdpi.com/article/10.3390/f12091274/s1, Table S1: Main ecological structure in the two municipalities. Table S2: Ecosystem Services identification and prioritization.

Author Contributions: Conceptualization A.S. and C.R.-A.; methodology, A.S. and E.C.-E.; validation, C.R.-A., E.C.-E., G.Y.F.-Y. and L.A.V.-M.; formal analysis, A.S.; investigation, A.S.; resources, A.S.; data curation, A.S. writing—original draft preparation, A.S.; writing—review and editing, C.R.-A., E.C.-E.; G.Y.F.-Y. and L.A.V.-M. All authors have read and agreed to the published version of the manuscript.

Funding: The authors thanks the Corporación Autónoma Regional de Caldas-CORPOCALDAS-for funding project number [222/2019], entitled 'Prospectar la biodiversidad y sus servicios ecosistémicos en la subregión norte del departamento de Caldas, como insumo técnico y jurídico para la determinante de estructura ecológica principal de la subregión Norte.' Furthermore, the authors acknowledge the Wildlife Conservation Society of Colombia for supporting and funding the consultancy process. Special thanks go to the young researchers who helped throughout the fieldwork process. Thanks to F. Gaviria for his support.

Institutional Review Board Statement: The study was conducted according to the guidelines of the Declaration of Helsinki and approved by the Institutional Review Board (or Ethics Committee) of Universidad de la Costa (Acta 88 de 2020).

Informed Consent Statement: Informed consent was obtained from all subjects involved in the study.

Data Availability Statement: Not applicable.

Conflicts of Interest: The authors declare no conflict of interest.

References

1. Fisher, B.; Turner, R.K.; Morling, P. Defining and classifying ecosystem services for decision making. *Ecol. Econ.* **2009**, *68*, 643–653. [CrossRef]
2. Haines-Young, R.; Potschin, M.B. *Common International Classification of Ecosystem Services (CICES) V5.1 and Guidance on the Application of the Revised Structure*; Fabis Consulting Ltd.: Nottingham, UK, 2018.
3. Maes, J.; Egoh, B.; Willemen, L.; Liquete, C.; Vihervaara, P.; Schägner, J.P.; Grizzetti, B.; Drakou, E.G.; La Notte, A.; Zulian, G.; et al. Mapping ecosystem services for policy support and decision making in the European Union. *Ecosyst. Serv.* **2012**, *1*, 31–39. [CrossRef]
4. Kandziora, M.; Burkhard, B.; Müller, F. Interactions of ecosystem properties, ecosystem integrity and ecosystem service indicators: A theoretical matrix exercise. *Ecol. Indic.* **2013**, *28*, 54–78. [CrossRef]
5. Wallace, K.J. Classification of ecosystem services: Problems and solutions. *Biol. Conserv.* **2007**, *139*, 235–246. [CrossRef]
6. Charles, M.; Ziv, G.; Bohrer, G.; Bakshi, B.R. Connecting air quality regulating ecosystem services with beneficiaries through quantitative serviceshed analysis. *Ecosyst. Serv.* **2019**, *41*, 101057. [CrossRef]
7. Escobedo, F.J.; Kroeger, T.; Wagner, J.E. Urban forests and pollution mitigation: Analyzing ecosystem services and disservices. *Environ. Pollut.* **2011**, *159*, 2078–2087. [CrossRef] [PubMed]
8. Higuera, D.; Martín-López, B.; Sánchez-Jabba, A. Social preferences towards ecosystem services provided by cloud forests in the neotropics: Implications for conservation strategies. *Reg. Environ. Chang.* **2012**, *13*, 861–872. [CrossRef]
9. Thompson, I.; Mackey, B.; McNulty, S.; Mosseler, A. Forest resilience, biodiversity, and climate change. A synthesis of the biodiversity/resilience/stability relationship in forest ecosystems. In *Secretariat of the Convention on Biological Diversity*; Technical Series No. 43; Secretariat of the Convention on Biological Diversity: Montreal, QC, Canada, 2009.
10. Baró, F.; Calderón-Argelich, A.; Langemeyer, J.; Connolly, J.J. Under one canopy? Assessing the distributional environmental justice implications of street tree benefits in Barcelona. *Environ. Sci. Policy* **2019**, *102*, 54–64. [CrossRef] [PubMed]
11. Feng, Z.; Cui, Y.; Zhang, H.; Gao, Y. Assessment of human consumption of ecosystem services in China from 2000 to 2014 based on an ecosystem service footprint model. *Ecol. Indic.* **2018**, *94*, 468–481. [CrossRef]

12. Song, C.; Lee, W.K.; Choi, H.A.; Kim, J.; Jeon, S.W.; Kim, J.S. Spatial assessment of ecosystem functions and services for air purification of forests in South Korea. *Environ. Sci. Policy* **2016**, *63*, 27–34. [CrossRef]
13. Baró, F.; Palomo, I.; Zulian, G.; Vizcaino, P.; Haase, D.; Gómez-Baggethun, E. Mapping ecosystem service capacity, flow and demand for landscape and urban planning: A case study in the Barcelona metropolitan region. *Land Use Policy* **2016**, *57*, 405–417. [CrossRef]
14. Xing, Y.; Brimblecombe, P. Role of vegetation in deposition and dispersion of air pollution in urban parks. *Atmos. Environ.* **2018**, *201*, 73–83. [CrossRef]
15. Yang, S.; Zhao, W.; Pereira, P.; Liu, Y. Socio-cultural valuation of rural and urban perception on ecosystem services and human well-being in Yanhe watershed of China. *J. Environ. Manag.* **2019**, *251*, 109615. [CrossRef] [PubMed]
16. Jeanloz, S.; Lizin, S.; Beenaerts, N.; Brouwer, R.; Van Passel, S.; Witters, N. Towards a more structured selection process for attributes and levels in choice experiments: A study in a Belgian protected area. *Ecosyst. Serv.* **2016**, *18*, 45–57. [CrossRef]
17. Zhang, H.; Pang, Q.; Long, H.; Zhu, H.; Gao, X.; Li, X.; Jiang, X.; Liu, K. Local Residents' Perceptions for Ecosystem Services: A Case Study of Fenghe River Watershed. *Int. J. Environ. Res. Public Health* **2019**, *16*, 3602. [CrossRef]
18. Baró, F.; Haase, D.; Gómez-Baggethun, E.; Frantzeskaki, N. Mismatches between ecosystem services supply and demand in urban areas: A quantitative assessment in five European cities. *Ecol. Indic.* **2015**, *55*, 146–158. [CrossRef]
19. Kibria, A.S.; Behie, A.; Costanza, R.; Groves, C.; Farrell, T. The value of ecosystem services obtained from the protected forest of Cambodia: The case of Veun Sai-Siem Pang National Park. *Ecosyst. Serv.* **2017**, *26*, 27–36. [CrossRef]
20. Lee, H.J.; Yoo, S.H.; Huh, S.Y. Economic benefits of introducing LNG-fuelled ships for imported flour in South Korea. *Transp. Res. Part D Transp. Environ.* **2020**, *78*, 102220. [CrossRef]
21. Shannon, A.K.; Usmani, F.; Pattanayak, S.K.; Jeuland, M. The Price of Purity: Willingness to pay for air and water purification technologies in Rajasthan, India. *Environ. Resour. Econ.* **2018**, *73*, 1073–1100. [CrossRef]
22. Li, T.; Gao, X. Ecosystem services valuation of lakeside wetland park beside Chaohu Lake in China. *Water* **2016**, *8*, 301. [CrossRef]
23. Pérez-Sánchez, D.; Montes, M.; Cardona-Almeida, C.; Vargas-Marín Luís, A.; Enríquez-Acevedo, T.; Suarez, A. Keeping people in the loop: Socioeconomic valuation of dry forest ecosystem services in the Colombian Caribbean region. *J. Arid Environ.* **2021**, *188*, 104446. [CrossRef]
24. Freeman, R.; Liang, W.; Song, R.; Timmins, C. Willingness to pay for clean air in China. *J. Environ. Econ. Manag.* **2019**, *94*, 188–216. [CrossRef]
25. Pu, S.; Shao, Z.; Yang, L.; Liu, R.; Bi, J.; Ma, Z. How much will the Chinese public pay for air pollution mitigation? A nationwide empirical study based on a willingness-to-pay scenario and air purifier costs. *J. Clean. Prod.* **2019**, *218*, 51–60. [CrossRef]
26. Wang, B.; Hong, G.; Qin, T.; Fan, W.R.; Yuan, X.C. Factors governing the willingness to pay for air pollution treatment: A case study in the Beijing-Tianjin-Hebei region. *J. Clean. Prod.* **2019**, *235*, 1304–1314. [CrossRef]
27. Yao, L.; Deng, J.; Johnston, R.J.; Khan, I.; Zhao, M. Evaluating willingness to pay for the temporal distribution of different air quality improvements: Is China's clean air target adequate to ensure welfare maximization? *Can. J. Agric. Econ.* **2019**, *67*, 215–232. [CrossRef]
28. Ligus, M. Measuring the Willingness to Pay for Improved Air Quality: A Contingent Valuation Survey. *Pol. J. Environ. Stud.* **2018**, *27*, 763–771. [CrossRef]
29. Soo, J.S.T. Valuing Air Quality in Indonesia Using Households' Locational Choices. Environ. *Environ. Resour. Econ.* **2017**, *71*, 755–776.
30. Ives, C.D.; Kendal, D. The role of social values in the management of ecological systems. *J. Environ. Manag.* **2014**, *144*, 67–72. [CrossRef]
31. Jones, N.A.; Shaw, S.; Ross, H.; Witt, K. Pinner, B. The study of human values in understanding and managing socialecological systems. *Ecol. Soc.* **2016**, *21*, 15. [CrossRef]
32. Pascual, U.; Balvanera, P.; Díaz, S.; Pataki, G.; Roth, E.; Stenseke, M.; Watson, R.T.; Dessane, E.B.; Islar, M.; Kelemen, E.; et al. Valuing nature's contributions to people: The IPBES approach. *Curr. Opin. Environ. Sustain.* **2017**, *26–27*, 7–16. [CrossRef]
33. Arias-Arévalo, P.; Gómez-Baggethun, E.; Martín-López, B.; Pérez-Rincón, M. Widening the evaluative space for ecosystem services: A taxonomy of plural values and valuation methods. *Environ. Values* **2018**, *27*, 29–53. [CrossRef]
34. Gómez-Baggethun, E.; Martín-López, B. Ecological economics perspectives on ecosystem services valuation. In *Handbook of Ecological Economics*; Martínez-Alier, J., Muradian, R., Eds.; Edward Elgar: Cheltenham, UK, 2019; pp. 260–282.
35. Jacobs, S.; Zafra-Calvo, N.; Gonzalez-Jimenez, D.; Guibrunet, L.; Benessaiah, K.; Berghöfer, A.; Chaves-Chaparro, J.; Díaz, S.; Gomez-Baggethun, E.; Lele, S.; et al. Use your power for good: Plural valuation of nature—The Oaxaca statement. *Glob. Sustain.* **2020**, *3*, e8. [CrossRef]
36. Kenter, J.O. Editorial: Shared, plural and cultural values. *Ecosyst. Serv.* **2016**, *21*, 175–183. [CrossRef]
37. Livia, W.P.; Pérez, J.J. Valoración económica de la calidad de aire y su impacto en registros epoc de Bucaramanga. *Aibi Rev. Investig. Adm. Ing.* **2014**, *2*, 13–18. [CrossRef]
38. Vieira, J.; Matos, P.; Mexia, T.; Silva, P.; Lopes, N.; Freitas, C.; Correia, O.; Santos-Reis, M.; Branquinho, C.; Pinho, P. Green spaces are not all the same for the provision of air purification and climate regulation services: The case of urban parks. *Environ. Res.* **2018**, *160*, 306–313. [CrossRef]

39. Pascual, U.; Muradian, R.; Brander, L.; Gómez-Baggethun, E.; Martín-López, B.; Verma, M.; Armsworth, P.; Christie, M.; Cornelissen, H.; Eppink, F.; et al. Chapter 5. The economics of valuing ecosystem services and biodiversity. In *The Economics of Ecosystems and Biodiversity. Ecological and Economic Foundations*; 2010. Available online: http://africa.teebweb.org/wp-content/uploads/2013/04/D0-Chapter-5-The-economics-of-valuing-ecosystem-services-and-biodiversity.pdf (accessed on 13 October 2020).
40. Pearce, D.; Atkinson, G.; Mourato, S. *Cost-Benefit Analysis and the Environment: Recent Developments*; Organisation for Economic Co-Operation and Development. Available online: https://www.oecd.org/governance/cost-benefit-analysis-and-the-environment-9789264085169-en.htm (accessed on 29 June 2020).
41. Da Motta, R.S.; Ortiz, R.A. Costs and Perceptions Conditioning Willingness to Accept Payments for Ecosystem Services in a Brazilian Case. *Ecol. Econ.* **2018**, *147*, 333–342. [CrossRef]
42. Chu, X.; Zhan, J.; Wang, C.; Hameeda, S.; Wang, X. Households' Willingness to Accept Improved Ecosystem Services and Influencing Factors: Application of Contingent Valuation Method in Bashang Plateau, Hebei Province, China. *J. Environ. Manag.* **2020**, *255*, 109925. [CrossRef] [PubMed]
43. Abdullah, S.; Jeanty, P.W. Willingness to pay for renewable energy: Evidence from a contingent valuation survey in Kenya. *Renew. Sustain. Energy Rev.* **2011**, *15*, 2974–2983. [CrossRef]
44. SIRAP Eje Cafetero. Clasificación de Ecosistemas Naturales Terrestres del eje Cafetero. 2013. Available online: https://www.wwf.org.co/?213162/Clasificacion-de-Ecosistemas-Naturales-Terrestres-del-Eje-Cafetero (accessed on 12 December 2020).
45. Maskey, B.; Singh, M. Households' willingness to pay for improved waste collection service in Gorkha municipality of Nepal. *Environments* **2017**, *4*, 77. Available online: https://www.mdpi.com/2076--3298/4/4/77 (accessed on 23 April 2021). [CrossRef]
46. Johnston, R.J.; Boyle, K.J.; Adamowicz, W.; Bennett, J.; Brouwer, R.; Cameron, T.A.; Hanemann, W.M.; Hanley, N.; Ryan, M.; Scarpa, R.; et al. Contemporary guidance for stated preference studies. *J. Assoc. Environ. Resour. Econ.* **2017**, *4*, 319–405. [CrossRef]
47. Loomis, J.B. WAEA keynote address: Strategies for overcoming hypothetical bias in stated preference surveys. *J. Agric. Resour. Econ.* **2014**, *39*, 34–46.
48. Martín-López, B.; Iniesta-Arandia, I.; García-Llorente, M.; Palomo, I.; Casado-Arzuaga, I.; Del Amo, D.G.; Gómez-Baggethun, E.; Otero-Rozas, E.; Palacios-Agundez, I.; Willaarts, B.; et al. Uncovering ecosystem service bundles through social preferences. *PLoS ONE* **2012**, *7*, e38970. [CrossRef]
49. Kennedy, E. Comparing Valuation Methods for Ecosystem Services in Amstelland. Applying Ecosystem Service Valuation Methods to Evaluate Land-Use Changes. Bachelor's Thesis, Vrije University Amsterdam, 2014. Available online: https://spinlab.vu.nl/wp-content/uploads/2016/09/Comparing_valuation_methods_for_ecosystem_services_in_Amstelland_BSCThesis_Eric_Kennedy_thesis.pdf (accessed on 24 January 2021).
50. Lo, A.Y.; Jim, C.Y. Willingness of residents to pay and motives for conservation of urban green spaces in the compact city of Hong Kong. *Urban For. Urban Green.* **2010**, *9*, 113–120. [CrossRef]
51. Logar, I.; van den Bergh, J.C. Respondent uncertainty in contingent valuation of preventing beach erosion: An analysis with a polychotomous choice question. *J. Environ. Manag.* **2012**, *113*, 184–193. [CrossRef] [PubMed]
52. Enriquez-Acevedo, T.; Botero, C.M.; Cantero-Rodelo, R.; Pertuz, A.; Suarez, A. Willingness to pay for Beach Ecosystem Services: The case study of three Colombian beaches. *Ocean Coast. Manag.* **2018**, *161*, 96–104. [CrossRef]
53. Stephenson, K.; Shabman, L. Does ecosystem valuation contribute to ecosystem decision making? Evidence from hydropower licensing. *Ecol. Econ.* **2019**, *163*, 1–8. [CrossRef]
54. Gómez-Baggethun, E.; Ruiz-Pérez, M. Economic valuation and the commodification of ecosystem services. *Prog. Phys. Geogr. Earth Environ.* **2011**, *35*, 613–628. [CrossRef]
55. Liu, S.; Costanza, R.; Farber, S.; Troy, A. Valuing ecosystem services. Theory practice, and need for transdisciplinary synthesis. *Ann. N. Y. Acad. Sci.* **2010**, *1185*, 54–78. [CrossRef]
56. Arias-Arévalo, P.; Martín-López, B.; Gómez-Baggethun, E. Exploring intrinsic, instrumental, and relational values for sustainable management of social-ecological systems. *Ecol. Soc.* **2017**, *22*, 43. [CrossRef]
57. Bretschger, L.; Pittel, K. Twenty Key Challenges in Environmental and Resource. *Environ. Resour. Econ.* **2020**, *77*, 725–750. [CrossRef]
58. Polasky, S.C.L.; Kling, S.A.; Levin, S.R.; Carpenter, G.C.; Daily, P.R.; Ehrlich, G.M.; Heal; Lubchenco, J.; Strassheim, H.; Beck, S. *Handbook of Behavioural Change and Public Policy*; Handbooks of Research on Public Policy Series; Edward Elgar: Cheltenham, UK, 2019.
59. Tadaki, M.; Sinner, J.; Chan, K.M.A. Making sense of environmental values: A typology of concepts. *Ecol. Soc.* **2017**, *22*, 7. [CrossRef]
60. Jacobs, S.; Martin-Lopez, B.; Barton, D.N.; Dunford, R.; Harrison, P.A.; Kelemen, E.; Saarikoski, H.; Termansen, M.; García-Llorente, M.; Gómez-Baggethun, E.; et al. The means determine the end—Pursuing integrated valuation in practice. *Ecosyst. Serv.* **2018**, *29*, 515–528. [CrossRef]
61. Laurans, Y.; Rankovic, A.; Billé, R.; Pirard, R.; Mermet, L. Use of ecosystem services economic valuation for decision making: Questioning a literature blindspot. *J. Environ. Manag.* **2013**, *119*, 208–219. [CrossRef]
62. Jacobs, S.; Dendoncker, N.; Martin-Lopez, B.; Barton, D.N.; Gomez-Baggethun, E.; Boeraeve, F.; McGrath, F.L.; Vierikko, K.; Geneletti, D.; Sevecke, K.J.; et al. A new valuation school: Integrating diverse values of nature in resource and land use decisions. *Ecosyst. Serv.* **2016**, *22*, 213–220. [CrossRef]

63. Brei, M.; Pérez-Barahona, A.; Strobl, E. Protecting species through legislation: The case of sea turtles. *Am. J. Agric. Econ.* **2020**, *102*, 300–328. [CrossRef]
64. Antoci, A.; Borghesi, S.; Russu, P. Don't feed the bears! Environmental defense expenditures and species-typical behaviour in an optimal growth model. *Macroecon. Dyn.* **2019**, *25*, 733–752. [CrossRef]
65. Drupp, M.A. Limits to substitution between ecosystem services and manufactured goods and implications for social discounting. *Environ. Resour. Econ.* **2018**, *69*, 135–158. [CrossRef]
66. Enríquez-Acevedo, T.; Pérez-Torres, J.; Ruiz-Agudelo, C.; Suarez, A. Seed dispersal by fruit bats in Colombia generates ecosystem services. *Agron. Sustain. Dev.* **2021**, *40*, 1–15. [CrossRef]
67. Tengö, M.; Brondizio, E.S.; Elmqvist, T.; Malmer, P.; Spierenburg, M. Connecting diverse knowledge systems for enhanced ecosystem governance: The multiple evidence base approach. *Ambio* **2014**, *43*, 579–591. [CrossRef] [PubMed]
68. Hill, R.; Breslow, S.; Le Buhn, G.; Quezada-Euánn, J.J.; Kwapong, P.; Nates-Parra, G.; Buchori, D.; Howlett, B.; Maués, M.M.; Saeed, S.; et al. Biocultural diversity, pollinators and their socio-cultural values. In *The Assessment Report of the Intergovernmental Science-Policy Platform on Biodiversity and Ecosystem Services on Pollinators, Pollination and Food Production*; Secretariat of the Intergovernmental Science-Policy Platform on Biodiversity and Ecosystem Services; Potts, S.G., Imperatriz-Fonseca, V.L., Ngo, H.T., Eds.; 2016; pp. 276–359. Available online: https://ainfo.cnptia.embrapa.br/digital/bitstream/item/157539/1/Chapter5-Pollination-Published.pdf (accessed on 16 September 2021).

forests

MDPI

Article

The Trade-Offs and Synergies of Ecosystem Services in Jiulianshan National Nature Reserve in Jiangxi Province, China

Jiayuan Feng [1], Fusheng Chen [1], Fangran Tang [1], Fangchao Wang [1,2], Kuan Liang [1], Lingyun He [1] and Chao Huang [1,*]

1 Key Laboratory of National Forestry and Grassland Administration on Forest Ecosystem Protection and Restoration of Poyang Lake Watershed, College of Forestry, Jiangxi Agricultural University, Nanchang 330045, China; fjyjiayuan@163.com (J.F.); chenfush@hotmail.com (F.C.); tangfangran0023@163.com (F.T.); wfangch@163.com (F.W.); liangkuan25@163.com (K.L.); helingyun1999@163.com (L.H.)

2 School of Management, Nanchang University, Nanchang 330045, China

* Correspondence: heipichao85@hotmail.com

Citation: Feng, J.; Chen, F.; Tang, F.; Wang, F.; Liang, K.; He, L.; Huang, C. The Trade-Offs and Synergies of Ecosystem Services in Jiulianshan National Nature Reserve in Jiangxi Province, China. *Forests* **2022**, *13*, 416. https://doi.org/10.3390/f13030416

Academic Editors: Elisabetta Salvatori and Giacomo Pallante

Received: 21 January 2022
Accepted: 3 March 2022
Published: 5 March 2022

Publisher's Note: MDPI stays neutral with regard to jurisdictional claims in published maps and institutional affiliations.

Abstract: Ecosystem services are directly related to human well-being. Previous studies showed that management policies and human activities alter the trade-offs and synergies of ecosystem services. Taking effective measures to manage the trade-offs and synergies of ecosystem services is essential to sustain ecological security and achieve a "win-win" situation between society and ecosystems. This study investigated the spatiotemporal changes of water yield, soil conservation, and carbon sequestration in the Jiulianshan National Nature Reserve from 2000 to 2020 based on the InVEST model. We distinguished spatial patterns of trade-offs and synergies between ecosystem services using the correlation relationship analysis. Then we analyzed the response of ecosystem services relationships among different vegetation types and elevation bands. The results showed that water yield and carbon sequestration presented an overall upward trend, while soil conservation remained a marginal degradation. Rising ecosystem services were mainly in the central, western, and southeastern regions, and declining areas were mainly distributed in the midwestern and northeastern fringes. Synergies spatially dominated the interactions among water yield, soil conservation, and carbon sequestration, and the trade-offs were primarily concentrated in the northern, southern, and southwestern fringes. Among the different vegetation types, synergies dominated ecosystem services in broad-leaved forests, coniferous forests, mixed forests, and Moso bamboo forests and in grass. The trade-offs were gradually reduced with elevation. This study highlighted that trade-off of ecosystem services should be incorporated into ecological management policies, strengthening the effectiveness of nature reserves in protecting and improving China's ecosystem services.

Keywords: ecosystem services; national nature reserve; spatiotemporal dynamics; trade-off; synergy

1. Introduction

Ecosystem services are defined as benefits derived from ecosystems and which are the basis for human survival and sustainable socioeconomic development [1]. However, the supply capacity of ecosystem services has decreased with global change and rapid socioeconomic development [2,3]. According to a report of the Millennium Ecosystem Assessment (MEA), approximately 60% of the world's ecosystem services are in degradation, which could threaten global ecological security [4]. Therefore, policymakers should better understand the spatiotemporal dynamics and relationships of ecosystem services for designing sustainable ecological management policies [5,6].

The contradiction between human demands and ecosystem diversity leads to the complex relationships of ecosystem services. Synergies and trade-offs are typical relationships of ecosystem services [7]. Synergies are situations in which the combined effect of both services is greater than the sum of their separate effects [8]. Trade-offs are situations in which one ecosystem service increases due to a decline in another ecosystem service [9]. In trade-off situations, management decisions need to be made between options for enhancing ecological benefits that cannot be satisfied simultaneously, leading to changes in ecosystem service interactions [10]. Reducing the impacts of trade-offs and enhancing synergies should be a key consideration in making ecosystem management policies [11]. Therefore, clarifying the trade-offs and synergies among multiple ecosystem services is necessary to design ecological protection and sustainable development policies [12].

Statistical analysis methods have been widely used to identify trade-offs and synergy relationships between ecosystem services in previous studies [13,14]. For example, Bai et al. applied correlation analysis to compare the relationships among ecosystem services in the Baiyangdian watershed [15]. Turner et al. used cluster analysis to identify ecosystem service bundle types, then analyzed the multiple interactions of ecosystem services in Denmark [16]. However, the relationships of ecosystem services are not always linear and vary over time and space [17]. Existing methods for ecosystem services trade-offs cannot reflect the non-linear relationships and spatial heterogeneity of ecosystem services. Moreover, most previous studies focused on quantifying ecosystem service relationships through the snapshot approach or static analysis [18,19]. Research for long-time series based on spatially explicit methods is needed to improve the efficiency of regional ecological management [20].

Conservation areas were designated to protect the environment and maintain ecological security [21]. Nature reserves are the main protected areas and effective measures for conserving ecosystem services [22]. Nature reserves have a higher supply capacity of ecosystem services than other regions [23]. However, the ecological environment in China has undergone tremendous pressure due to rapid socioeconomic transformations [24]. Nature reserves will face an unprecedented challenge for maintaining long-term sustainable development within this context. This requires policymakers to formulate policies that minimize trade-offs of ecosystem services and maximize win-win relationships between society and ecosystems in nature reserves [25]. Therefore, it is necessary to identify the dynamics and relationships of ecosystem services in nature reserves to provide valid ecological management proposals [26]. This will promote the effectiveness of nature reserves in protecting the ecological environment and implementing sustainable development goals.

Nature reserves could benefit from the balance between strict protection and sustainable use of natural resources, contributing significantly to the ecological security of human society [27]. Although nature reserves are generally believed to be the cornerstones of ecosystem conservation and the safest strongholds of wildlife, the problems of biodiversity degradation, mismanagement, and human encroachments are still widespread [28–30]. Jiulianshan National Nature Reserve is a typical forest reserve located at a transitional belt between middle and south subtropical evergreen forests in China [31]. This reserve is an important ecological barrier area in southeastern China [32]. It is necessary to identify relationships of ecosystem services in nature reserves to assess the existing ecological management strategies. This study aims to provide a theoretical basis for evaluating the effectiveness of nature reserves on sustainable development and effective policy support on the ecological management of nature reserves. We investigated their spatial and temporal changes in water yield, soil conservation, and carbon sequestration from 2000 to 2020 based on InVEST model. We identified the trade-offs and synergies of three ecosystem services through correlation analysis. Then we analyzed the differences in trade-offs and synergies of ecosystem services among different vegetation types and elevation bands. Specifically, we (1) examined the spatiotemporal patterns of ecosystem services under the existing ecological management policies, (2) analyzed the spatial characteristics of trade-offs and synergies of ecosystem services, and (3) quantified the differences in trade-offs and synergies of ecosystem services among different vegetation types and elevation bands.

2. Materials and Methods

2.1. Study Area

Jiulianshan National Nature Reserve (24°29′18″–24°38′55″ N, 114°22′50″–114°31′32″ E) is located in southern Jiangxi Province, China (Figure 1). The total area of this reserve is 13,411.6 ha, spanning almost 15.7 km north-south and 15 km east-west. The elevation ranged from 280 to 1434 m, and the terrain is low in the north and high in the south. Jiulianshan National Nature Reserve belongs to the East Asian monsoon climate zone, with a warm and humid climate and distinct dry and wet seasons. The average annual precipitation is 2155.6 mm, the average annual temperature is 16.4 °C, and the annual evaporation is 790.2 mm. Jiulianshan National Nature Reserve is highly valuable in biodiversity and has a typical, large-scale and well-preserved subtropical forest ecosystem. The main type of vegetation in this area is the broad-leaved evergreen forest, including *Castanopsis carlesii* (Hemsl., Hayata.), *Schima superba* (Gardn. et Champ.), and *Litsea elongata* (Wall. ex Nees Benth. et Hook. F). Coniferous forests are also widespread, such as *Pinus massoniana* (Lamb.), *Cunninghamia lanceolata* (Lamb., Hook)., and *Taxus wallichiana var. mairei* (Lemee & H. Léveillé, L. K. Fu & Nan Li). Moso bamboo forests are mainly composed of *Phyllostachys edulis* (Carriere, J. Houzeau) (Supplementary Material Figure S1). In addition to vegetation resources there are abundant animal resources, such as 384 wild vertebrates and 1404 insect species. Moreover, there are 9 rural settlements in reserve, and most of the residents are engaged in agriculture and forestry production.

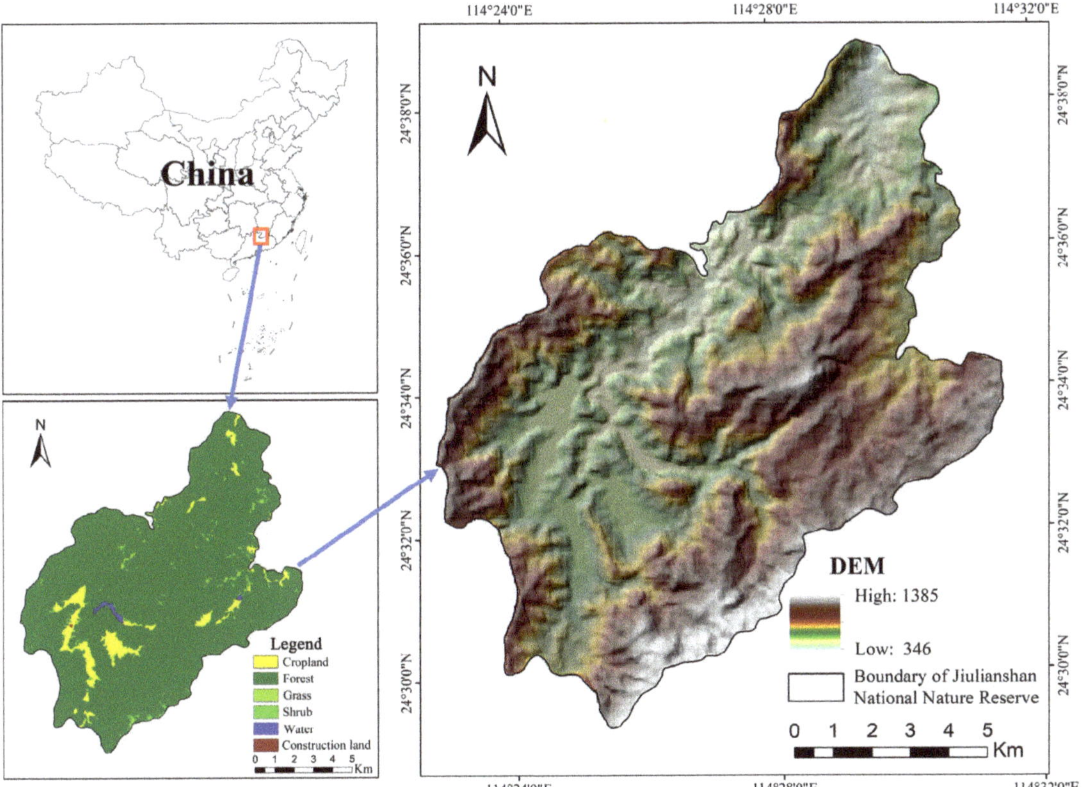

Figure 1. The geographic location, elevation, and land use/land cover types of the study area.

2.2. Methods

2.2.1. Model Parameterization

The land use/land cover map for 2000, 2010, and 2020 was obtained from the 30 m resolution Global Land Cover Dataset (GlobeLand30, http://www.globallandcover.com/, accessed on 15 January 2022). The 30 m Digital Elevation Model (DEM) data were provided by the Chinese Academy of Sciences (http://www.gscloud.cn, accessed on 15 January 2022). Soil data were provided by the Harmonized World Soil Database (Harmonized World Soil Database, HWSD). Meteorological data, including precipitation, temperature, and radiation, were obtained from the Jiulianshan National Nature Reserve Administration and the National Meteorological Administration of China (http://data.cma.cn, accessed on accessed on 15 January). The soil erodibility factor (K) was provided by the Center for Geodata and Analysis, Faculty of Geographical Science, Beijing Normal University (https://gda.bnu.edu.cn/, accessed on 15 January 2022) [33]. For model parameterization, all data were resampled to a 30 m grid.

The potential evapotranspiration (ET_0) was calculated based on the Modified-Hargreaves model [34]. Wischmeier's monthly scale formula was used to calculate the rainfall erosivity factor (R) [35]. The input raster maps (temperature, precipitation, ET_0, and R) were interpolated to a 30 m resolution by the Machine Learning Ensemble & Thin-Plate-Spline Interpolation model (MACHISPLIN). Aboveground carbon stocks and soil carbon stocks were collected from forest inventory data and previous studies in our study area (Supplementary Material Table S1) [36–38]. The root depth data and other parameters in the biophysical table were set according to the recommendations of the InVEST 3.8.9 model user's guide and published research articles (Supplementary Material Tables S2 and S3) [39–41].

2.2.2. Ecosystem Services Assessment

The Integrated Valuation of Ecosystem Service and Tradeoffs (InVEST) was used to evaluate ecosystem services from 2000 to 2020. InVEST model is a spatially explicit model for assessing ecosystem services under different land use/land cover types or different socioeconomic scenarios. This model explores how changes in ecosystems benefit people [42]. It has been widely used in ecosystem service research due to its advantages of low model parameters, low data requirements, and global applicability [43]. Water yield, soil conservation, and carbon sequestration were simulated using InVEST 3.8.9 at a 30 m spatial resolution.

(1) Water yield

The water yield model of the InVEST model was mainly based on the Budyko curve and the annual average precipitation. Water yield is calculated by subtracting the actual evaporation from the precipitation of each grid cell [44]. Water yield is calculated as follows:

$$Y_{xj} = \left(1 - \frac{E_{xj}}{P_x}\right) \cdot P_x \tag{1}$$

where Y_{xj} is the water yield of land use/land cover type j in pixel x (mm/year), E_{xj} is the annual actual evapotranspiration of land use/land cover type j in pixel x (mm/year), and P_x is the average annual precipitation in pixel x (mm/year).

(2) Soil conservation

The sediment delivery ratio sub-model of the InVEST model can be used to calculate the potential soil erosion (RKLS), actual soil erosion (USLE), and soil conservation (SD) for each cell as follows:

$$RKLS = R \times K \times LS, \tag{2}$$

$$USLE = R \times K \times LS \times P \times C, \tag{3}$$

$$SD = RKLS - USLE, \tag{4}$$

where RKLS, USLE, and SD represent potential soil erosion [t/(ha·year)], actual soil erosion [t/(ha·year)], and soil conservation [t/(ha·year)], respectively. R is the rainfall erosivity index [MJ·mm/(ha·h·year)]. K is the soil erodibility index [t·ha·h/(MJ·ha·mm)]. LS is the length-gradient factor. C is the cover-management factor for the USLE, and P is the supporting practice factor for the USLE.

(3) Carbon sequestration

The carbon storage and sequestration model of the InVEST model takes the different land use/land cover types as the evaluation unit, then estimates the biomass of each carbon pool by multiplying the average carbon density of the different land use/land cover types by the area of each evaluation unit. The total carbon sequestration of terrestrial ecosystems is the sum of four fundamental carbon pools, including the carbon sequestration of aboveground biomass, belowground biomass, soil organic matter, and dead organic matter. The total carbon sequestration is calculated as follows:

$$C_{total} = C_{above} + C_{below} + C_{soil} + C_{dead}, \quad (5)$$

where C_{total} is the total carbon stocks (t/ha), C_{above} is aboveground carbon stocks (t/ha), C_{below} is belowground carbon stocks (t/ha), C_{soil} is soil organic carbon stocks (t/ha), and C_{dead} is dead organic matter carbon stocks (t/ha).

2.2.3. Model Calibration and Validation

The seasonal parameter Z-value is an empirical constant that indicates the regional precipitation distribution and geohydrological characteristics [45]. Correlation between simulated data and field survey data can ensure the accuracy of the water yield model. In this study, the average runoff volume (4.55×10^6 m^3) of the Jiulianshan National Nature Reserve No.4 measuring weir (24°31′45″ N, 114°27′36″ E) in 2010 was used as a reference to calibrate the accuracy of the simulation results. The results of the correlation analysis indicated that the water yield was negatively correlated with the Z-value ($R^2 = 0.75$, $p < 0.01$), and the fitted equation showed that the simulated water yield matches best with the natural runoff in 2010 when Z = 3.33 (Figure 2).

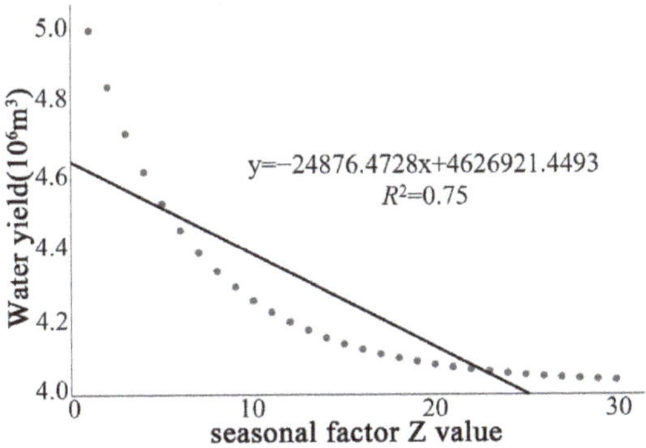

Figure 2. The correlation between water source content and seasonal factor Z value.

We selected 100 forest survey plots (20 m × 20 m) to validate the carbon storage and sequestration model of the InVEST model. At each plot we identified species and measured the diameter at breast height of trees over 1 cm. Three randomly placed sample squares (5 m × 5 m) within each sample plot were used for soil sampling. The results showed that the simulation results were close to the forest survey data and that the correlation coefficient between the two is 0.78, which means a high correlation ($R^2 = 0.78$, $p < 0.01$, Figure 3).

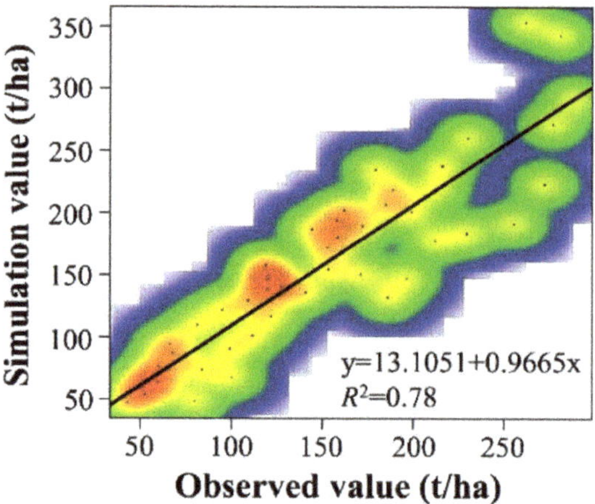

Figure 3. Comparison of observed and simulated carbon storage from the InVEST model. The observed carbon storage was estimated by forest inventory data conducted in 2020.

2.2.4. Analysis of Trade-Offs and Synergies of Ecosystem Services

Correlation relationship analysis was applied to analyze trade-offs and synergies of ecosystem services. When the correlation coefficient between two ecosystem services is positive, they are considered synergistic, and when the correlation coefficient is negative, it is a trade-off relationship [46]. The correlation coefficients were calculated as follows:

$$r_{12(ij)} = \frac{\sum_1^n \left(ES1_{n(ij)} - \overline{ES1_{n(ij)}}\right) \times \sum_1^n \left(ES2_{n(ij)} - \overline{ES2_{n(ij)}}\right)}{\sqrt{\sum_1^n \left(ES1_{n(ij)} - \overline{ES1_{n(ij)}}\right)^2 \times \sum_1^n \left(ES2_{n(ij)} - \overline{ES2_{n(ij)}}\right)^2}} \quad (6)$$

where ES1 and ES2 are two ecosystem services; r is the correlation coefficient between ES1 and ES2; i and j are the number of rows and columns of raster image elements; n is the time series of raster data; and $r_{12(ij)}$ is the correlation coefficient between ES1 and ES2 on image element ij when other ecosystem services change at n years. $r_{12(ij)} > 0$ indicates a synergistic relationship between ES1 and ES2; $r_{12(ij)} = 0$ indicates no correlation between ES1 and ES2; $r_{12(ij)} < 0$ indicates trade-off relationship between ES1 and ES2; and the larger value of $r_{12(ij)}$ indicates a stronger correlation between ES1 and ES2.

We classified the correlation coefficients into seven levels: no relationship (r = 0); synergy** (r > 0, 0.01 < p < 0.05); synergy* (r > 0, 0.05 < p < 0.1); synergy (r > 0, 0.1 < p); trade-off (r < 0, 0.01 < p); trade-off* (r < 0, 0.05 < p < 0.1); and trade-off** (r < 0, 0.01 < p < 0.05, Table 1).

Table 1. The relevance classification level of trade-off and synergy relationships between ecosystem services.

Relevance Classification Level	r	p
No relationship	r = 0	0.01 < p < 0.05
Synergy **	r > 0	0.01 < p < 0.05
Synergy *	r > 0	0.05 < p < 0.10
Synergy	r > 0	0.10 < p
Trade-off	r < 0	0.10 < p
Trade-off *	r < 0	0.05 < p < 0.10
Trade-off **	r < 0	0.01 < p < 0.05

*: 0.05 < p < 0.10; **: 0.01 < p < 0.05.

3. Results

3.1. Changes in Ecosystem Services from 2000 to 2020

3.1.1. The Spatial Pattern of Ecosystem Services

Water yield, soil conservation, and carbon sequestration maintained a stable spatial pattern in 2000–2020. Three ecosystem services exhibited their spatial distribution characteristics (Figure 4). The higher value of the water yield was clumped within the central, northern, and northwestern areas (Figure 4a), where there were high forest cover and precipitation. In contrast, the lower value of water yield was mainly concentrated in the central-western and southeastern areas. The areas with a lower value of water yield in 2020 decreased in the southeast compared to 2000 and 2010. The higher-value areas of soil conservation were scattered; they were mainly distributed in the study area's central, western, and southeastern areas (Figure 4b). The spatial pattern of carbon sequestration has not changed substantially because of the stable land use/land cover structure. The areas with a higher value of carbon sequestration are mainly concentrated in forests, while the areas with a lower value were in cropland, in water, and on construction land (Figure 4c).

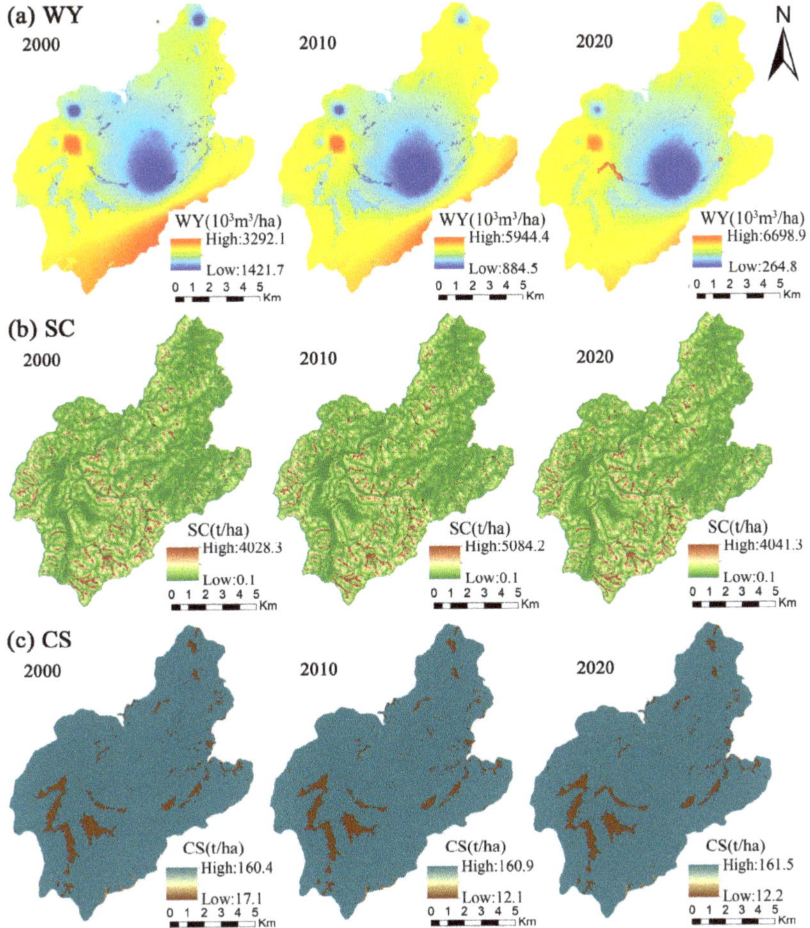

Figure 4. The spatial pattern of ecosystem services in the study area in 2000, 2010, and 2020; (**a**) water yield; (**b**) soil conservation; (**c**) carbon sequestration.

3.1.2. Temporal and Spatial Changes in Ecosystem Services from 2000 to 2020

The changes in water yield and carbon sequestration exhibited an upward trend from 2000 to 2020, while soil conservation remained a marginal degradation. Water yield increased by 13.4% and carbon sequestration increased by 0.4% in 2000–2020. Soil conservation declined marginally from 14.85×10^6 t in 2000 to 14.55×10^6 t in 2020. The areas where the water yield increased were mainly concentrated in the central areas (Figure 5a). The spatial distribution of the soil conservation rising areas was different. From 2000 to 2010, the rising soil conservation areas were mainly concentrated in the central, western, and southern areas, while from 2010 to 2020 they were mainly clumped within the eastern, northeastern, and southwestern regions (Figure 5b). The supply capacity of carbon sequestration was slightly enhanced in the central, northeastern, and southwestern areas (Figure 5c).

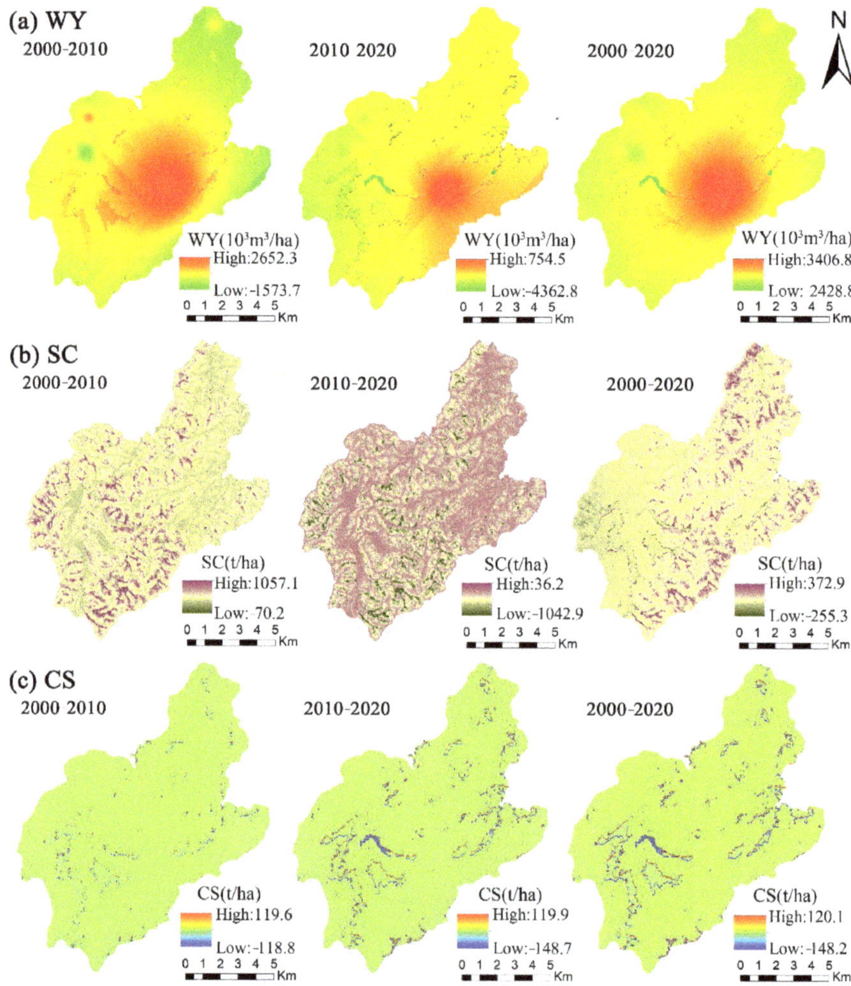

Figure 5. Spatial and temporal variation in ecosystem services in the study area from 2000 to 2010, 2010 to 2020, and 2000 to 2020; (**a**) water yield; (**b**) soil conservation; (**c**) carbon sequestration.

3.2. Trade-Offs and Synergies of Ecosystem Services

We calculated the pixel-scale spatial correlation coefficients to explore the trade-offs and synergies between each ecosystem service (Figure 6). The result showed that synergies were spatially dominated in water yield, soil conservation, and carbon sequestration interactions. The synergies areas accounted for 67.90% of soil conservation and water yield and were spatially aggregated in the central and southern regions. The weak trade-offs were mainly concentrated in the eastern, northern, and western areas, while strong trade-offs were concentrated in the northern, southern, and southwestern marginal regions (Figure 6a). For soil conservation and carbon sequestration, the synergistic relationships were mainly distributed in the study areas' central, southern, and western parts, which accounted for 90.35%. The trade-off areas were small and dispersed and were spatially aggregated in the northern, southern, and southwestern fringes (Figure 6b). For carbon sequestration and water yield, the synergistic relationships were mainly clumped within the eastern, central, and southern areas, which accounted for 89.14%. The trade-off areas were spatially aggregated in the northern and western regions (Figure 6c).

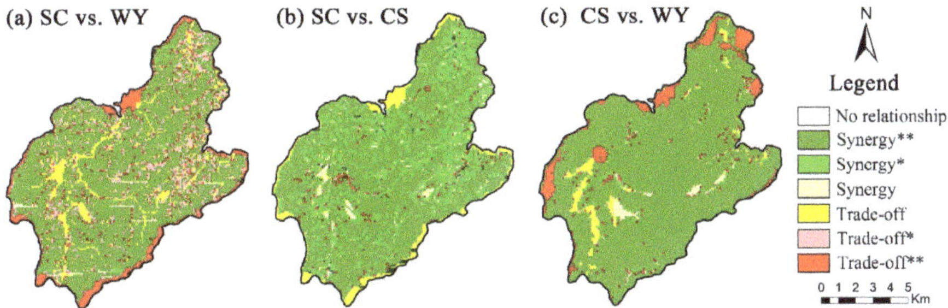

Figure 6. The spatial pattern of trade-off and synergy among three ecosystem services; (**a**) soil conservation and water yield; (**b**) soil conservation and carbon sequestration; (**c**) carbon sequestration and water yield. *: $0.05 < p < 0.10$; **: $0.01 < p < 0.05$.

The distribution of trade-offs and synergies among ecosystem services in various vegetation types is shown in Figure 7. Synergistic relationships between soil conservation and carbon sequestration were dominant in cropland, grass, and shrub, and occupied 84.89%, 77.45%, and 51.29%, respectively (Figure 7a). The trade-off relationships were most significantly concentrated in broad-leaved, coniferous, mixed, and Moso bamboo forests. The trade-offs between soil conservation and water yield were mostly concentrated in cropland and shrub, which accounted for 70.25% and 69.24% (Figure 7b). The synergistic relationships were mostly in broad-leaved forests, coniferous forests, mixed forests, and Moso bamboo forests and in grass. For carbon sequestration and water yield, the synergies were significantly concentrated in various vegetation types, which all accounted for more than 50% (Figure 7c).

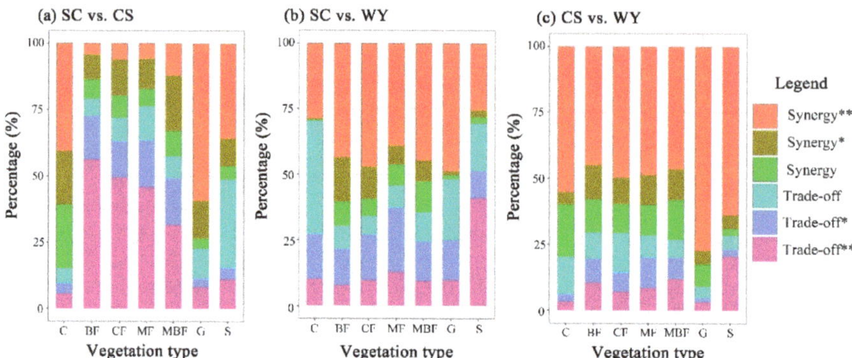

Figure 7. Trade-offs and synergistic of ecosystem services among different vegetation types. C, cropland; BF, broad-leaved forests; CF, coniferous forests; MF, mixed forests; MBF, Moso bamboo forests; G, grass; S, shrub; (**a**) soil conservation and water yield; (**b**) soil conservation and carbon sequestration; (**c**) carbon sequestration and water yield. *: $0.05 < p < 0.10$; **: $0.01 < p < 0.05$.

Figure 8 shows the distribution of ecosystem services of trade-off and synergies situations at different elevation bands. For soil conservation and carbon sequestration the trade-offs were concentrated in all elevation bands, and they all accounted for more than 50% (Figure 8a). The trade-offs of areas below 1000 m were significantly stronger than that above 1000 m. Synergies were mostly concentrated in all elevation bands (Figure 8b). Among them, strong synergies were dominant. For soil conservation and water yield the synergistic relationships were concentrated in areas above 500 m, and the trade-offs were mainly in areas below 500 m (Figure 8c). The proportion of trade-offs increased with elevation.

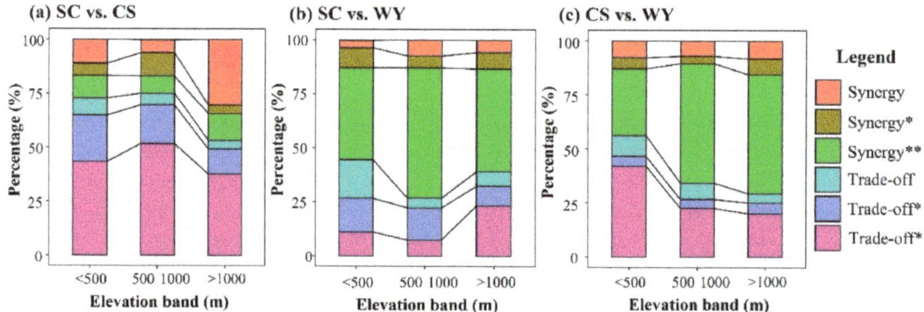

Figure 8. Trade-offs and synergistic of ecosystem services at different elevation bands; (**a**) soil conservation and water yield; (**b**) soil conservation and carbon sequestration; (**c**) carbon sequestration and water yield. *: $0.05 < p < 0.10$; **: $0.01 < p < 0.05$.

4. Discussion

4.1. Spatial and Temporal Variation in Ecosystem Services

In this study we found that the supply capacity of water yield and carbon sequestration exhibited an overall upward trend from 2000 to 2020, while soil conservation remained a marginal degradation. This result indicates that ecosystem services in Jiulianshan National Nature Reserve presented improvements over the 20 years. However, there were significant differences in the spatial distribution of variation in ecosystem services (Figure 5). The spatial and temporal changes of ecosystem services could be associated with climate [47]. Previous research has shown that elevated temperature and precipitation within a certain range are conducive to ecosystem services [48,49]. The warm and humid climate in the conservation area could favor vegetation improvement. Compared with low vegetation cover areas, high vegetation cover areas have a stronger ability to enhance water yield and control soil erosion by intercepting rainfall, increasing infiltration, and stabilizing the soil [50]. In addition, implementing ecological restoration projects positively impacts ecosystem services, which could protect the ecological environment by restoring and protecting forests [51]. However, human activities, such as excessive land utilization, could decrease supply to multiple ecosystem services in partial areas [52]. In the conservation area, the low-altitude and gentle terrain of the midwestern and northeastern regions could facilitate agricultural production activities by the residents. Without appropriate compensatory measures, agricultural activities may cause a reduction in soil fertility and change the structure of the soil [53]. To better harmonize the relationship between human activities and the natural environment, the midwestern and northeastern areas of the conservation area should receive increased attention.

The results suggest that the existing protection policies of the Jiulianshan National Nature Reserve effectively curbed deforestation and land reclamation, but more detailed management measures are still needed. The reserve should also further increase financial investment and cooperate with social welfare organizations to obtain more operating funds. The financial investment in protected areas indirectly reflects ecosystem management policy and positively relates to ecosystem service conservation effectiveness [54]. Meanwhile, realizing the diversification of residents' income sources is essential to reducing the destruction of natural resources and improving ecosystem services' supply capacity [55]. Therefore, decision-makers should also consider the spatial heterogeneity of ecosystem services and their natural and environmental drivers.

4.2. Quantifying Trade-Offs and Synergistic Relationships of Ecosystem Services

Our research showed the dominant synergies between water yield, soil conservation, and carbon sequestration in the Jiulianshan National Nature Reserve. The results are generally consistent with those of previous studies [56–58]. Nearly 95% of the conservation area consists of forests, grasslands, and shrub. The high vegetation cover may explain the strong synergies of ecosystem services, as it enhances carbon sequestration by photosynthesis and mitigates soil erosion [59]. The supply capacity of the water yield could be enhanced, because vegetation intercepts, absorbs, and stores precipitation from the canopy of dead leaf vegetation and the subsoil [51,60]. The distribution of trade-offs in the study area was dispersed and concentrated in the north, midwestern, and southeastern areas, especially in peripheral areas. This may be interpreted as the consequence of the effects of human activities. Uncontrolled development of resources could cause artificial surface expansion, and land cover/land use types or vegetation types may change accordingly [61]. The central areas of the conservation area are densely vegetated, and land-use conflicts were weak there. Under strict ecological policies, the major areas were relatively less affected by external drivers, which may explain why there were strong synergies between ecosystem services in these areas. When making management policies, more consideration should be given to the spatial heterogeneity of ecosystem services, which will benefit the refinement of ecological management measures.

This reserve has a typical and well-preserved subtropical forest system. In the broad-leaved, coniferous, mixed, and Moso bamboo forests, synergies among water yield and other services were dominant. Forests are the main ecological construction and restoration means, by providing several intangible benefits such as regulating air humidity, protecting watersheds, and absorbing carbon and nutrient cycling [62]. However, forests cannot simultaneously produce multiple ecosystem services due to the trade-offs among competing functions. Forests could absorb carbon by photosynthesis and enhance the supply capacity of carbon sequestration, but excessive rainfall intensity could lead to soil erosion. These may result in trade-offs between soil conservation and carbon sequestration. Therefore, the construction of soil and water conservation engineering measures should be strengthened in forestlands with high precipitation and rainfall intensity. Trade-offs between soil conservation and water yield were primarily concentrated in cropland and shrub. Water-saving and pollution reduction measures should be implemented in cropland, such as drip irrigation and water and fertilizer integration [50]. Moreover, the reserve should take some comprehensive protection measures in the shrub, such as increasing vegetation density and improving vegetation structure to strengthen the plant root system and upgrade water retention [63]. In addition, we found that trade-offs among ecosystem services decreased with increasing elevation bands. In the low elevation zone, policymakers should pay more attention to trade-offs between soil conservation and carbon sequestration. Commercial plants could be incorporated into mixed farming, to protect vegetation diversity and the livelihood flexibility of farmers in reserve [64]. In the areas above 1000 m, the forests are primeval and well-preserved. Enhancing patrolling efforts is efficient to reduce human interference and prevent further damage [65]. Moreover, the Jiulianshan National Nature Reserve should be managed more scientifically and constructed with corresponding engineering measures [66]. Consider establishing coordination mechanisms with multiple stakeholders when implementing ecological restoration measures. In addition to these macro-protection principles, we should also consider the suitability of various vegetation types for different environments and elevations [67].

4.3. *Limitations*

The InVEST model was used to quantitatively evaluate water yield, soil conservation, and carbon sequestration from 2000 to 2020. This model has been recognized as an appropriate tool because it achieves the expected results with fewer data and simplified computational steps [43]. Our study improved the accuracy and reliability of the evaluation results through model validation. The validation results showed that the study results were in good agreement with the actual value. However, the modeling and data limitations of the InVEST model should be recognized. This model cannot address daily or monthly climate data, so our study ignored daily or monthly extreme climatic events [68]. Moreover, the carbon-storage-and-sequestration model uses a simplified carbon cycle that quantifies the amount of static carbon storage, which lacks consideration of the transformation between various carbon pools. Therefore, extreme climate effects on ecosystem services and the optimization of the accuracy of simulation models need more attention in future research to reduce the uncertainty of model assessment results.

5. Conclusions

This study investigated spatial and temporal changes in water yield, soil conservation, and carbon sequestration from 2000 to 2020 in the Jiulianshan Nation Nature Reserve by applying the InVEST model. Spatial patterns of trade-offs and synergies were distinguished by correlation analysis. Then we analyzed the response of ecosystem services relationships among different vegetation types and elevation bands. Our results showed that water yield and carbon sequestration exhibited an overall upward trend from 2000 to 2020, while soil conservation remained a marginal degradation. Ecosystem services increased, in aggregate, in the central, western, and southeastern areas, and decreased in the midwestern areas and northeastern fringes. The synergies among water yield, soil conservation, and carbon sequestration were spatially dominant, while the trade-offs were mainly concentrated in the northern, southern, and southwestern areas. Among the different vegetation types, synergies dominated ecosystem services in broad-leaved forests, coniferous forests, mixed forests, and Moso bamboo forests and in grass. Moreover, with the increase in elevation, the trade-offs were gradually reduced. Therefore, to maximize the benefits of ecosystem services, policymakers should focus on the trade-off areas and strengthen the implementation of ecological projects based on the balance of ecosystem services. This study could provide a theoretical basis for implementing management strategies and the sustainable development of society, economy, and ecology in China's nature reserves.

Supplementary Materials: The following supporting information can be downloaded at: https://www.mdpi.com/article/10.3390/f13030416/s1, Figure S1: The vegetation and land use/land cover type of the study area; Table S1: Input data on carbon stored in each of the four fundamental pools for each LULC class in the InVEST 3.8.9 model (t/ha); Table S2: Input data for each LULC class in the InVEST 3.8.0 water yield model; Table S3: Input data for each LULC class in the InVEST 3.8.9 sediment delivery ratio model.

Author Contributions: Conceptualization, J.F., F.C., F.T. and C.H.; Methodology, J.F., F.T., K.L., L.H., F.W. and C.H.; Validation, J.F., F.T., K.L., F.W. and C.H.; Data curation and analysis, J.F., F.C., F.T., F.W. and C.H.; Writing-Review & Editing, J.F., F.C., F.T., K.L., L.H., F.W. and C.H. All authors have read and agreed to the published version of the manuscript.

Funding: This research was funded by the National Natural Science Foundation of China (32160292 and 31800408), the National Key R&D Program of China (2017YFA0604403), and the Double Thousand Plan of Jiangxi Province (jxsq2020101080).

Institutional Review Board Statement: Not applicable.

Informed Consent Statement: Not applicable.

Data Availability Statement: Land use/land cover data used in this study is available at http://www.globallandcover.com (accessed on 15 January 2022). The 30 m Digital Elevation Model (DEM) data is available in http://www.gscloud.cn (accessed on 15 January 2022). The soil erodibility factor (K) is available in https://gda.bnu.edu.cn/(accessed on 15 January 2022).

Acknowledgments: We would like to thank the high-performance computing support from the Center for Geodata and Analysis, Faculty of Geographical Science, Beijing Normal University (https://gda.bnu.edu.cn/, accessed on 15 January 2022).

Conflicts of Interest: The authors declare no conflict of interest.

References

1. Costanza, R.; d'Arge, R.; De Groot, R.; Farber, S.; Grasso, M.; Hannon, B.; Limburg, K.; Naeem, S.; O'Neill, R.V.; Paruelo, J.; et al. The value of the world's ecosystem services and natural capital. *Nature* **1997**, *25*, 3–15. [CrossRef]
2. Ouyang, Z.; Zheng, H.; Xiao, Y.; Polasky, S.; Liu, J.; Xu, W.; Wang, Q.; Zhang, L.; Xiao, Y.; Rao, E.; et al. Improvements in ecosystem services from investments in natural capital. *Science* **2016**, *352*, 1455–1459. [CrossRef] [PubMed]
3. Balkanlou, K.R.; Müller, B.; Cord, A.F.; Panahi, F.; Malekian, A.; Jafari, M.; Egli, L. Spatiotemporal dynamics of ecosystem services provision in a degraded ecosystem: A systematic assessment in the Lake Urmia basin, Iran. *Sci. Total Environ.* **2020**, *716*, 137100. [CrossRef] [PubMed]
4. Scholes, R.J.; Ash, N.; Hassan, R.M. *Ecosystems and Human Well-Being*; Island Press: Washington, DC, USA, 2005.
5. Luty, L.; Musiał, K.; Zioło, M. The Role of Selected Ecosystem Services in Different Farming Systems in Poland Regarding the Differentiation of Agricultural Land Structure. *Sustainability* **2021**, *13*, 6673. [CrossRef]
6. Assumma, V.; Bottero, M.; Caprioli, C.; Datola, G.; Mondini, G. Evaluation of Ecosystem Services in Mining Basins: An Application in the Piedmont Region (Italy). *Sustainability* **2022**, *14*, 872. [CrossRef]
7. Cord, A.F.; Bartkowski, B.; Beckmann, M.; Dittrich, A.; Hermans-Neumann, K.; Kaim, A.; Lienhoop, N.; Locher-Krause, K.; Priess, J.; Schröter-Schlaack, C.; et al. Towards systematic analyses of ecosystem service trade-offs and synergies: Main concepts, methods and the road ahead. *Ecosyst. Serv.* **2017**, *28*, 264–272. [CrossRef]
8. Haase, D.; Schwarz, N.; Strohbach, M.; Kroll, F.; Seppelt, R. Synergies, Trade-offs, and Losses of Ecosystem Services in Urban Regions: An Integrated Multiscale Framework Applied to the Leipzig-Halle Region, Germany. *Ecol. Soc.* **2012**, *17*, 22. [CrossRef]
9. Tomscha, S.A.; Gergel, S.E. Ecosystem service trade-offs and synergies misunderstood without landscape history. *Ecol. Soc.* **2016**, *21*, 43. [CrossRef]
10. Deng, X.; Li, Z.; Gibson, J. A review on trade-off analysis of ecosystem services for sustainable land-use management. *J. Geogr. Sci.* **2016**, *26*, 953–968. [CrossRef]
11. Felipe-Lucia, M.R.; Comín, F.A.; Bennett, E.M. Interactions Among Ecosystem Services Across Land Uses in a Floodplain Agroecosystem. *Ecol. Soc.* **2014**, *19*, 20. [CrossRef]
12. Yang, Y.; Zheng, H.; Kong, L.; Huang, B.; Xu, W.; Ouyang, Z. Mapping ecosystem services bundles to detect high- and low-value ecosystem services areas for land use management. *J. Clean. Prod.* **2019**, *225*, 11–17. [CrossRef]
13. Howe, C.; Suich, H.; Vira, B.; Mace, G.M. Creating win-wins from trade-offs? Ecosystem services for human well-being: A meta-analysis of ecosystem service trade-offs and synergies in the real world. *Glob. Environ. Change* **2014**, *28*, 263–275. [CrossRef]
14. Han, Z.; Song, W.; Deng, X.; Xu, X. Trade-Offs and Synergies in Ecosystem Service within the Three-Rivers Headwater Region, China. *Water* **2017**, *9*, 588. [CrossRef]

15. Bai, Y.; Zhuang, C.; Ouyang, Z.; Zheng, H.; Jiang, B. Spatial characteristics between biodiversity and ecosystem services in a human-dominated watershed. *Ecol. Complex.* **2011**, *8*, 177–183. [CrossRef]
16. Turner, K.G.; Odgaard, M.V.; Bøcher, P.K.; Dalgaard, T.; Svenning, J. Bundling ecosystem services in Denmark: Trade-offs and synergies in a cultural landscape. *Landsc. Urban Plan.* **2014**, *125*, 89–104. [CrossRef]
17. Ndong, G.O.; Therond, O.; Cousin, I. Analysis of relationships between ecosystem services: A generic classification and review of the literature. *Ecosyst. Serv.* **2020**, *43*, 101120. [CrossRef]
18. Yang, S.; Zhao, W.; Liu, Y.; Wang, S.; Wang, J.; Zhai, R. Influence of land use change on the ecosystem service trade-offs in the ecological restoration area: Dynamics and scenarios in the Yanhe watershed, China. *Sci. Total Environ.* **2018**, *644*, 556–566. [CrossRef]
19. Fu, B.J.; Yu, D.D. Trade-off analyses and synthetic integrated method of multiple ecosystem services. *Resour. Sci.* **2016**, *38*, 1–9.
20. Jopke, C.; Kreyling, J.; Maes, J.; Koellner, T. Interactions among ecosystem services across Europe: Bagplots and cumulative correlation coefficients reveal synergies, trade-offs, and regional patterns. *Ecol. Indic.* **2015**, *49*, 46–52. [CrossRef]
21. Huang, Y.; Fu, J.; Wang, W.; Li, J. Development of China's nature reserves over the past 60 years: An overview. *Land Use Policy* **2019**, *80*, 224–232. [CrossRef]
22. Xua, X.; Jiang, B.; Chen, M.; Bai, Y.; Yang, G. Strengthening the effectiveness of nature reserves in representing ecosystem services: The Yangtze River Economic Belt in China. *Land Use Policy* **2020**, *96*, 104717. [CrossRef]
23. Scolozzi, R.; Schirpke, U.; Morri, E.; D'Amato, D.; Santolini, R. Ecosystem services-based SWOT analysis of protected areas for conservation strategies. *J. Environ. Manag.* **2014**, *146*, 543–551. [CrossRef] [PubMed]
24. Liu, Y.L.J.; Yang, Y. Strategic adjustment of land use policy under the economic transformation. *Land Use Policy* **2018**, *74*, 5–14. [CrossRef]
25. Figgis, P.; Mackey, B.; Fitzsimons, J.; Irving, J.; Clarke, P. *Valuing Nature: Protected Areas and Ecosystem Services*; Australian Committee for IUCN: Sydney, Australia, 2015.
26. Schirpke, U.; Marino, D.; Marucci, A.; Palmieri, M.; Scolozzi, R. Operationalising ecosystem services for effective management of protected areas: Experiences and challenges. *Ecosyst. Serv.* **2017**, *28*, 105–114. [CrossRef]
27. Xu, W.; Xiao, Y.; Zhang, J.; Yang, W.; Zhang, L.; Hull, V.; Wang, Z.; Zheng, H.; Liu, J.; Polasky, S.; et al. Strengthening protected areas for biodiversity and ecosystem services in China. *Proc. Natl. Acad. Sci. USA* **2017**, *114*, 1601–1606. [CrossRef]
28. Liu, J.; Linderman, M.; Ouyang, Z.; An, L.; Yang, J.; Zhang, H. Ecological degradation in protected areas: The case of Wolong Nature Reserve for giant pandas. *Science* **2001**, *292*, 98–101. [CrossRef]
29. Pouzols, F.M.; Toivonen, T.; Di Minin, E.; Kukkala, A.S.; Kullberg, P.; Kuusterä, J.; Lehtomäki, J.; Tenkanen, H.; Verburg, P.H.; Moilanen, A. Global protected area expansion is compromised by projected land-use and parochialism. *Nature* **2014**, *516*, 383–386. [CrossRef]
30. Wu, J.; Yang, J.; Nakagoshi, N.; Lu, X.; Xu, H. Proposal for Developing Carbon Sequestration Economy in Ecological Conservation Area of Beijing. *Adv. Mater. Res.* **2012**, *524*, 3424–3427.
31. Jian, M.-F.; Liu, Q.-J.; Zhu, D.; You, H. Inter-specific correlations among dominant populations of tree layer species in evergreen broad-leaved forest in Jiulianshan Mountain of subtropical China. *Chin. J. Plant Ecol.* **2009**, *33*, 672–680.
32. Li, H. Valuation of the Ecosystem Service Functions of Jiangxi Jiulianshan National Nature Reserve. *For. Resour. Manag.* **2006**, *4*, 70–73.
33. Yue, T.; Yin, S.; Xie, Y.; Yu, B.; Liu, B. Rainfall erosivity mapping over mainland China based on high density hourly rainfall records. *Earth Syst. Sci. Data Discuss.* **2022**, *14*, 665–682. [CrossRef]
34. Allen, R.G.; Pereira, L.S.; Raes, D.; Smith, M. Crop evapotranspiration-Guidelines for computing crop water requirements-FAO Irrigation and drainage paper 56. *Fao Rome* **1998**, *300*, D05109.
35. Smith, D.D.; Wischmeier, W.H. Predicting Rainfall Erosion Losses: A Guide to Conservation Planning. Department of Agriculture, Science and Education Administration. 1978. Available online: https://naldc.nal.usda.gov/catalog/CAT79706928 (accessed on 15 January 2022).
36. Cancan, Z.; Xiaogang, W.; Bin, L.; Xuewen, S.; Fusheng, C.; Lihong, Q.; Wensheng, B. Variations in soil organic carbon along an altitudinal gradient of Jiulian Mountain in Jiangxi Province of eastern China. *J. Beijing For. Univ.* **2019**, *41*, 19–28.
37. Liu, Z.; Dong, X.; Liu, Z. Organic Carbon Storage and Spatial Distribution of Forest Soil in Jiangxi. *Adv. Mater. Res.* **2014**, *1010*, 1194–1197.
38. Li, H.; Wang, S.; Gao, L.; Yu, G. The carbon storage of the subtropical forest vegetation in central Jiangxi Province. *Acta Ecol. Sin.* **2007**, *27*, 693–702.
39. Liu, J.; Fu, B.; Zhang, C.H.; Peng, Z.H.; Wang, Y.K. Assessment of Ecosystem Water Retention and Its Value in the Upper Reaches of Minjiang River Based on InVEST Model. *Resour. Environ. Yangtze Basin* **2019**, *28*, 578–585.
40. Redhead, J.W.; Stratford, C.; Sharps, K.; Jones, L.; Ziv, G.; Clarke, D.; Oliver, T.H.; Bullock, J.M. Empirical validation of the InVEST water yield ecosystem service model at a national scale. *Sci. Total Environ.* **2016**, *569*, 1418–1426. [CrossRef]
41. Pan, T.; Wu, S.-H.; Dai, E.; Liu, Y.-J. Spatiotemporal variation of water source supply service in Three Rivers Source Area of China based on InVEST model. *Chin. J. Appl. Ecol.* **2013**, *24*, 183–189.
42. Sharp, R.; Douglass, J.; Wolny, S. *InVEST User's Guide*; The Natural Capital Project: Stanford, CA, USA, 2014.
43. Yang, D.; Liu, W.; Tang, L.; Chen, L.; Li, X.; Xu, X. Estimation of water provision service for monsoon catchments of South China: Applicability of the InVEST model. *Landsc. Urban Plan.* **2019**, *182*, 133–143. [CrossRef]

44. Feng, X.M.; Sun, G.; Fu, B.J.; Su, C.H.; Liu, Y.; Lamparski, H. Regional effects of vegetation restoration on water yield across the Loess Plateau, China. *Hydrol. Earth Syst. Sci.* **2012**, *16*, 2617–2628. [CrossRef]
45. Zhang, L.; Dawes, W.R.; Walker, G.R. Response of mean annual evapotranspiration to vegetation changes at catchment scale. *Water Resour. Res.* **2001**, *37*, 701–708. [CrossRef]
46. Li, T.; Lü, Y.; Fu, B.; Hu, W.; Comber, A.J. Bundling ecosystem services for detecting their interactions driven by large-scale vegetation restoration: Enhanced services while depressed synergies. *Ecol. Indic.* **2019**, *99*, 332–342. [CrossRef]
47. Li, Z.; Cheng, X.; Han, H. Future Impacts of Land Use Change on Ecosystem Services under Different Scenarios in the Ecological Conservation Area, Beijing, China. *Forests* **2020**, *11*, 584. [CrossRef]
48. Lang, Y.; Song, W.; Zhang, Y. Responses of the water-yield ecosystem service to climate and land use change in Sancha River Basin, China. *Phys. Chem. Earth Parts A/B/C* **2017**, *101*, 102–111. [CrossRef]
49. Wang, X.; Chu, B.; Feng, X.; Li, Y.; Fu, B.; Liu, S.; Jin, J. Spatiotemporal variation research on the driving factors of water yield services on the Qingzang Plateau. *Geogr. Sustain.* **2021**, *2*, 31–39.
50. Sun, X.; Lu, Z.; Li, F.; Crittenden, J.C. Analyzing spatio-temporal changes and trade-offs to support the supply of multiple ecosystem services in Beijing, China. *Ecol. Indic.* **2018**, *94*, 117–129. [CrossRef]
51. Huang, L.; Wang, B.; Niu, X.; Gao, P.; Song, Q. Changes in ecosystem services and an analysis of driving factors for China's Natural Forest Conservation Program. *Ecol. Evol.* **2019**, *9*, 3700–3716. [CrossRef]
52. Zhang, P.; He, L.; Fan, X.; Huo, P.; Liu, Y.; Zhang, T.; Pan, Y.; Yu, Z. Ecosystem Service Value Assessment and Contribution Factor Analysis of Land Use Change in Miyun County, China. *Sustainability* **2015**, *7*, 7333–7356. [CrossRef]
53. Power, A.G. Ecosystem services and agriculture: Tradeoffs and synergies. *Philos. Trans. R. Soc. B Biol. Sci.* **2010**, *365*, 2959–2971. [CrossRef]
54. Machado, M.; Young, C.E.F.; Clauzet, M. Environmental funds to support protected areas: Lessons from Brazilian experiences. *Parks* **2020**, *26*, 47. [CrossRef]
55. Qian, D.; Cao, G.; Du, Y.; Li, Q.; Guo, X. Impacts of climate change and human factors on land cover change in inland mountain protected areas: A case study of the Qilian Mountain National Nature Reserve in China. *Environ. Monit. Assess.* **2019**, *191*, 486. [CrossRef] [PubMed]
56. Crouzat, E.; Mouchet, M.; Turkelboom, F.; Byczek, C.; Meersmans, J.; Berger, F.; Verkerk, P.J.; Lavorel, S. Assessing bundles of ecosystem services from regional to landscape scale: Insights from the French Alps. *J. Appl. Ecol.* **2015**, *52*, 1145–1155. [CrossRef]
57. Zhang, X.; Xie, H.; Shi, J.; Lv, T.; Zhou, C.; Liu, W. Assessing Changes in Ecosystem Service Values in Response to Land Cover Dynamics in Jiangxi Province, China. *Int. J. Environ. Res. Public Health* **2020**, *17*, 3018. [CrossRef] [PubMed]
58. Yin, L.; Wang, X.; Zhang, K.; Xiao, F.; Cheng, C.; Zhang, X. Trade-offs and synergy between ecosystem services in National Barrier Zone. *Geogr. Res.* **2019**, *38*, 2162–2172.
59. Braun, D.; Damm, A.; Hein, L.; Petchey, O.L.; Schaepman, M.E. Spatio-temporal trends and trade-offs in ecosystem services: An Earth observation based assessment for Switzerland between 2004 and 2014. *Ecol. Indic.* **2018**, *89*, 828–839. [CrossRef]
60. Dai, E.-F.; Wang, X.-L.; Zhu, J.-J.; Xi, W.-M. Quantifying ecosystem service trade-offs for plantation forest management to benefit provisioning and regulating services. *Ecol. Evol.* **2017**, *7*, 7807–7821. [CrossRef]
61. Ran, C.; Wang, S.; Bai, X.; Tan, Q.; Zhao, C.; Luo, X.; Chen, H.; Xi, H. Trade-Offs and Synergies of Ecosystem Services in Southwestern China. *Environ. Eng. Sci.* **2020**, *37*, 669–678. [CrossRef]
62. Palomo, I. Climate Change Impacts on Ecosystem Services in High Mountain Areas: A Literature Review. *Mt. Res. Dev.* **2017**, *37*, 179–187. [CrossRef]
63. Keesstra, S.; Nunes, J.; Novara, A.; Finger, D.; Avelar, D.; Kalantari, Z.; Cerdà, A. The superior effect of nature based solutions in land management for enhancing ecosystem services. *Sci. Total Environ.* **2018**, *610*, 997–1009. [CrossRef]
64. Xu, J. China's new forests aren't as green as they seem. *Nature* **2011**, *477*, 371. [CrossRef]
65. Muttaqin, M.Z.; Alviya, I.; Lugina, M.; Hamdani, F.A.U. Indartik Developing community-based forest ecosystem service management to reduce emissions from deforestation and forest degradation. *For. Policy Econ.* **2019**, *108*, 101938. [CrossRef]
66. Wang, Y.; Li, X.; Zhang, Q.; Li, J.; Zhou, X. Projections of future land use changes: Multiple scenarios-based impacts analysis on ecosystem services for Wuhan city, China. *Ecol. Indic.* **2018**, *94*, 430–445. [CrossRef]
67. Miao, N.; Xu, H.; Moermond, T.C.; Li, Y.; Liu, S. Density-dependent and distance-dependent effects in a 60-ha tropical mountain rain forest in the Jianfengling Mountains, Hainan Island, China: Spatial pattern analysis. *For. Ecol. Manag.* **2018**, *429*, 226–232. [CrossRef]
68. Hamel, P.; Guswa, A.J. Uncertainty analysis of a spatially explicit annual water-balance model: Case study of the Cape Fear basin, North Carolina. *Hydrol. Earth Syst. Sci.* **2015**, *19*, 839–853. [CrossRef]

MDPI
St. Alban-Anlage 66
4052 Basel
Switzerland
Tel. +41 61 683 77 34
Fax +41 61 302 89 18
www.mdpi.com

Forests Editorial Office
E-mail: forests@mdpi.com
www.mdpi.com/journal/forests

www.ingramcontent.com/pod-product-compliance
Lightning Source LLC
LaVergne TN
LVHW070708170726
843515LV00004B/519

* 9 7 8 3 0 3 6 5 4 9 7 5 0 *